Lecture Notes in Computer Science 12448

More information about this series at http://www.springer.com/series/7407

Sylvain Schmitz · Igor Potapov (Eds.)

Reachability Problems

14th International Conference, RP 2020
Paris, France, October 19–21, 2020
Proceedings

 Springer

Editors
Sylvain Schmitz
Université de Paris
Paris, France

Igor Potapov
University of Liverpool
Liverpool, UK

ISSN 0302-9743   ISSN 1611-3349 (electronic)
Lecture Notes in Computer Science
ISBN 978-3-030-61738-7   ISBN 978-3-030-61739-4 (eBook)
https://doi.org/10.1007/978-3-030-61739-4

LNCS Sublibrary: SL1 – Theoretical Computer Science and General Issues

This Springer imprint is published by the registered company Springer Nature Switzerland AG
The registered company address is: Gewerbestrasse 11, 6330 Cham, Switzerland

Preface

This volume contains the papers presented at the 14th International Conference on Reachability Problems (RP 2020), organized by the University of Paris, France. Previous events in the series were located at: Université Libre de Bruxelles, Belgium (2019), Aix-Marseille University, France (2018), Royal Holloway, University of London, UK (2017), Aalborg University, Denmark (2016), the University of Warsaw, Poland (2015), the University of Oxford, UK (2014), Uppsala University, Sweden (2013), the University of Bordeaux, France (2012), the University of Genoa, Italy (2011), Masaryk University, Czech Republic (2010), École Polytechnique, France (2009), The University of Liverpool, UK (2008), and University of Turku, Finland (2007).

The aim of the conference is to bring together scholars from diverse fields with a shared interest in reachability problems, and to promote the exploration of new approaches for the modeling and analysis of computational processes by combining mathematical, algorithmic, and computational techniques. Topics of interest include (but are not limited to): reachability for infinite state systems; rewriting systems; reachability analysis in counter/timed/cellular/communicating automata; Petri nets; computational game theory, computational aspects of semigroups, groups, and rings; reachability in dynamical and hybrid systems; frontiers between decidable and undecidable reachability problems; complexity and decidability aspects; predictability in iterative maps; and new computational paradigms.

We are very grateful to our invited speakers, who gave the following talks:

- **Valérie Berthé**, University of Paris, IRIF, CNRS, France:
  "On Decision Problems for Substitutions in Symbolic Dynamics"
- **Patricia Bouyer-Decitre**, CNRS, ENS Paris-Saclay, France:
  "When are Finite Games Finite-Memory Determined?"
- **Kousha Etessami**, The University of Edinburgh, UK:
- "Computing a Fixed Point of a Monotone Function, and Some Applications"
- **Orna Kupferman**, The Hebrew University, Israel:
  "Games with Full, Longitudinal and Transverse Observability"
- **Dirk Nowotka**, Kiel University, Germany:
  "Word Equations: From Theory to Practice"

The conference received 25 submissions (17 regular and 8 presentation-only submissions) from which 3 regular papers were withdrawn. Each submission was carefully reviewed by three Program Committee (PC) members. Based on these reviews, the PC decided to accept 8 regular papers. The members of the PC and the list of external reviewers can be found on the next pages. We are grateful for the high quality work produced by the PC and the external reviewers. Overall this volume contains 8 contributed papers and 2 papers from invited speakers which cover their talks.

The conference also provided the opportunity for other young and established researchers to present work in progress or work already published elsewhere. This year, in addition to 8 regular submissions, the PC selected 8 high-quality presentations on various reachability aspects in theoretical computer science. A list of accepted talk-only submissions is given below:

PrIC3: Property Directed Reachability for MDPs

Kevin Batz, Sebastian Junges, Benjamin Lucien Kaminski, Joost-Pieter Katoen, Christoph Matheja, and Philipp Schröer

Abstract: IC3 has been a leap forward in symbolic model checking. This paper proposes PrIC3 (pronounced pricy-three), a conservative extension of IC3 to symbolic model checking of MDPs. Our main focus is to develop the theory underlying PrIC3. Alongside, we present a first implementation of PrIC3 including the key ingredients from IC3 such as generalization, repushing, and propagation. This paper has been published in the CAV 2020 proceedings.

Higher-Order Nonemptiness Step by Step

Paweł Parys

Abstract: We show a new simple algorithm that checks whether a given higher-order grammar generates a nonempty language of trees. The algorithm amounts to a procedure that transforms a grammar of order n to a grammar of order $n - 1$, preserving nonemptiness, and increasing the size only exponentially. After repeating the procedure n times, we obtain a grammar of order 0, whose nonemptiness can be easily checked. Since the size grows exponentially at each step, the overall complexity is n-EXPTIME, which is known to be optimal. More precisely, the transformation (and hence the whole algorithm) is linear in the size of the grammar, assuming that the arity of employed nonterminals is bounded by a constant. The same algorithm allows to check whether an infinite tree generated by a higher-order recursion scheme is accepted by an alternating safety (or reachability) automaton, because this question can be reduced to the nonemptiness problem by taking a product of the recursion scheme with the automaton. A proof of correctness of the algorithm is formalized in the proof assistant Coq. Our transformation is motivated by a similar transformation of Asada and Kobayashi (2020) changing a word grammar of order n to a tree grammar of order $n - 1$. The step-by-step approach can be opposed to previous algorithms solving the nonemptiness problem "in one step", being compulsorily more complicated. This paper was submitted to the FSTTCS 2020 conference.

The Strahler Number of a Parity Game

Laure Daviaud, Marcin Jurdzinski, and K. S. Thejaswini

Abstract: The Strahler number of a rooted tree is the largest height of a perfect binary tree that is its minor. The Strahler number of a parity game is proposed to be defined as the smallest Strahler number of the tree of any of its attractor decompositions. It has been proven that parity games can be solved in quasi-linear space and in time that is polynomial in the number of vertices n and linear in $(d/2k)^k$, where d is the number of priorities and k is the Strahler number. This complexity is quasi-polynomial because the Strahler number is at most logarithmic in the number of vertices. The proof is based on a new construction of small Strahler-universal trees. It is shown that the Strahler number of a parity game is robust, and hence arguably a natural, parameter: it coincides with its alternative version based on trees of progress measures – and remarkably – with the register number defined by Lehtinen (2018). It follows that parity games can be solved in quasi-linear space and in time that is polynomial in the number of vertices and linear in $(d/2k)^k$, where k is the register number. This significantly improves the running times and space achieved for parity games of bounded register number by Lehtinen (2018) and by Parys (2020). The running time of the algorithm based on small Strahler-universal trees yields a novel trade-off $k \cdot lg(d/k) = O(log(n))$ between the two natural parameters that measure the structural complexity of a parity game, which allows solving parity games in polynomial time. This includes special cases, for example the asymptotic settings of those parameters covered by the results of Calude, Jain, Khoussainov, Li, and Stephan (2017), of Jurdzinski and Lazic (2017), and of Lehtinen (2018), and it significantly extends the range of such settings, for example to $d = 2^{O(\sqrt{\lg n})}$ and $k = O(\sqrt{\lg n})$. This paper was published in the ICALP 2020 proceedings.

Cost Automata, Safe Schemes, and Downward Closures
David Barozzini, Lorenzo Clemente, Thomas Colcombet, and Paweł Parys
Abstract: Higher-order recursion schemes are an expressive formalism used to define languages of possibly infinite ranked trees. They extend regular and context-free grammars, and are equivalent to simply typed λY-calculus and collapsible pushdown automata. In this work we prove, under a syntactical constraint called safety, decidability of the model-checking problem for recursion schemes against properties defined by alternating B-automata, an extension of alternating parity automata for infinite trees with a boundedness acceptance condition. We then exploit this result to show how to compute downward closures of languages of finite trees recognized by safe recursion schemes. This paper was published in the ICALP 2020 proceedings.

Decidability of cutpoint isolation for probabilistic finite automata on letter-bounded inputs
Paul Bell and Pavel Semukhin
Abstract: We show the surprising result that the cutpoint isolation problem is decidable for probabilistic finite automata where input words are taken from a letter-bounded context-free language. A context-free language $\mathcal{L}$ is letter-bounded when $\mathcal{L} \subseteq a_1^* a_2^* \cdots a_\ell^*$ for some finite $\ell > 0$ where each letter is distinct. A cutpoint is isolated when it cannot be approached arbitrarily closely. The decidability of this problem is in marked contrast to the situation for the (strict) emptiness problem for PFA which is undecidable under the even more severe restrictions of PFA with polynomial ambiguity, commutative matrices, and input over a letter-bounded language, as well as to the injectivity problem which is undecidable for PFA over letter-bounded languages. We provide a constructive nondeterministic algorithm to solve the cutpoint isolation problem, which holds even when the PFA is exponentially ambiguous. We also show that the problem is at least NP-hard and use our decision procedure to solve several related problems. This paper was published in the CONCUR 2020 proceedings.

On Polynomial Recursive Sequences

Michaël Cadilhac, Filip Mazowiecki, Charles Paperman, Michał Pilipczuk, and Géraud Sénizergues

Abstract: We study the expressive power of polynomial recursive sequences, a non-linear extension of the well-known class of linear recursive sequences. These sequences arise naturally in the study of nonlinear extensions of weighted automata, where (non)expressiveness results translate to class separations. A typical example of a polynomial recursive sequence is $b_n = n!$. Our main result is that the sequence $u_n = n^n$ is not polynomial recursive. This work was published in the proceedings of ICALP 2020.

The Complexity of Reachability in Affine Vector Addition Systems with States

Mikhail Raskin and Michael Blondin

Abstract: Vector addition systems with states (VASS) are widely used for the formal verification of concurrent systems. Given their tremendous computational complexity, practical approaches have relied on techniques such as reachability relaxations, e.g., allowing for negative intermediate counter values. It is natural to question their feasibility for VASS enriched with primitives that typically translate into undecidability. Spurred by this concern, we pinpoint the complexity of integer relaxations w.r.t. arbitrary classes of affine operations. More specifically, we provide a trichotomy on the complexity of integer reachability in VASS extended with affine operations (affine VASS). Namely, we show that it is NP-complete for VASS with resets, PSPACE-complete for VASS with (pseudo-)transfers and VASS with (pseudo-)copies, and undecidable for any other class. We further present a dichotomy for standard reachability in affine VASS: it is decidable for VASS with permutations, and undecidable for any other class. This yields a complete and unified complexity landscape of reachability in affine VASS. This work has previously appeared in the proceedings of LICS 2020.

Additionally, we present our ongoing work on the possible complexity of reachability and integer reachability in specific affine VASS. We obtain for every nontrivial computable predicate p a corresponding A-VASS such that its integer reachability and reachability for that VASS are equivalent to p under 1-1 polynomial reductions. In the other words, unlike the case matrix classes specific instances can have virtually arbitrary complexity of reachability and integer reachability relations.

An Approach to Regular Separability in Vector Addition Systems

Wojciech Czerwiński and Georg Zetzsche

Abstract: We study the problem of regular separability of languages of vector addition systems with states (VASS). It asks whether for two given VASS languages K and L, there exists a regular language R that includes K and is disjoint from L. While decidability of the problem in full generality remains an open question, there are several subclasses for which decidability has been shown: It is decidable for (i) one-dimensional VASS, (ii) VASS coverability languages, (iii) languages of integer VASS, and (iv) commutative VASS languages. We propose a general approach to deciding regular separability. We use it to decide regular separability of an arbitrary VASS language from any language in the classes (i), (ii), and (iii). This generalizes all previous results, including (iv). This paper was published in the LICS 2020 proceedings.

So overall, the conference program consisted of five invited talks, eight presentations of contributed papers, and eight informal presentations in the area of reachability problems stretching from results on fundamental questions in mathematics and computer science to efficient solutions of practical problems.

It is a pleasure to thank the team behind the EasyChair system and the Lecture Notes in Computer Science team at Springer, who together made the production of this volume possible in time for the conference. Finally, we thank all the authors and invited speakers for their high-quality contributions, and the participants for making RP 2020 a success. We are also very grateful to Alfred Hofmann for the continuous support of the event in the last decade, to the Research Institute on the Foundations of Computer Science, and Springer for their financial sponsorship.

October 2020   Sylvain Schmitz
                                                                            Igor Potapov

Organization

Program Committee

C. Aiswarya	CMI, India
S. Akshay	IIT Bombay, India
Christel Baier	TU Dresden, Germany
Sergiy Bogomolov	University of Newcastle, UK
Olivier Bournez	École Polytechnique, France
Laure Daviaud	City University of London, UK
Pierre Ganty	IMDEA Software Institute, Spain
Gilles Geeraerts	Free University of Brussels, Belgium
Matthew Hague	Royal Holloway, University of London, UK
Mika Hirvensalo	Turku University, Finland
Petr Jančar	Palacký University Olomouc, Czech Republic
Raphaël Jungers	UCLouvain, Belgium
Akitoshi Kawamura	Kyoto University, Japan
Sang-Ki Ko	Kangwon National University, South Korea
Slawomir Lasota	University of Warsaw, Poland
Karoliina Lehtinen	The University of Liverpool, UK
Igor Potapov	The University of Liverpool, UK
Cristian Riveros	Pontifical Catholic University of Chile, Chile
Sylvain Schmitz	University of Paris, France
Mahsa Shirmohammadi	CNRS, France
Georg Zetzsche	MPI-SWS, Germany

Program Chairs

Igor Potapov	The University of Liverpool, UK
Sylvain Schmitz	University of Pairs, France

Additional Reviewers

Christoph Haase	Paulius Stankaitis
Kostiantyn Potomkin	Maciej Koutny
Faisal Khan	Pascal Baumann
Joel Day	Filip Mazowiecki
Patrick Totzke	James Worrell

Abstracts of Invited Talks

On Decision Problems for Substitutions
in Symbolic Dynamics

Valérie Berthé🄳

Université de Paris, IRIF, CNRS, F-75013 Paris, France
berthe@irif.fr

Abstract. In this survey, we discuss decidability issues for symbolic dynamical systems generated by substitutions. Symbolic dynamical systems are discrete dynamical systems made of infinite sequences of symbols, with the shift acting on them. Substitutions are simple rules that replace letters by string of letters and allow the generation of infinite words. We focus here on symbolic dynamical systems that are generated by infinite compositions of substitutions, allowing to go beyond the case of the iteration of a single substitution. This is the so-called *S*-adic framework. Motivated by decidability and ergodic questions, we focus on questions dealing with the convergence of products of nonnegative matrices and associated Lyapounov exponents.

Keywords: Substitutions · Symbolic dynamics · Decidability · Lyapunov exponents · Primitive matrix · Perron–Frobenius theorem

This work was supported by the Agence Nationale de la Recherche through the project Codys (ANR-18-CE40-0007).

When are Finite Games Finite-Memory Determined?

Patricia Bouyer

Université Paris-Saclay, CNRS, ENS Paris-Saclay, France
bouyer@lsv.fr

Abstract. In their CONCUR 2005 paper [1], Gimbert and Zielonka gave a complete characterization of preference relations for which memoryless optimal strategies always exist in finite turn-based (two-player antagonistic) games. As an important consequence of their characterization, they furthermore establish that two-player games are memoryless determined for a given preference relation if and only if their one-player counterparts are both memoryless determined. This is of utmost practical importance, since it allows to infer memoryless determinacy of two-player games by proving memoryless determinacy of one-player games (which is likely to be much easier).

Though memoryless strategies are the simplest ones and therefore much more desirable, more complex objectives often require memory (finite or infinite). In most cases, ad-hoc proofs are designed to analyze the required memory, and despite some effort, no such elegant characterization had been proposed for memoryless optimal strategies earlier.

We present here a complete characterization of preference relations for which *arena-independent* finite-memory optimal strategies always exist, which generalizes the work by Gimbert and Zielonka to the finite-memory case. This result enjoys the same important practical corollary as the memoryless case, which allows one to deduce finite-memory determinacy results of two-player games from finite-memory determinacy of one-player games. This characterization strictly generalizes the Gimbert-Zielonka characterization for memoryless optimal strategies, and covers the case of arena-independent memory (for instance for multiple parity objectives or for lower- and upper-bounded energy objectives). The setting of arena-dependent finite memory (as needed by multiple lower-bounded energy objectives) requires further investigation.

This talk is based on joint work with Stéphane Le Roux, Youssouf Oualhadj, Mickael Randour, and Pierre Vandenhove. The technical paper is published in the proceedings of CONCUR 2020 [2].

References

1. Gimbert, H., Zielonka, W.: Games where you can play optimally without any memory. In: Abadi, M., de Alfaro, L. (eds.) CONCUR 2005. LNCS, vol. 3653, pp. 428-442. Springer, Heidelberg (2005). https://doi.org/10.1007/11539452_33
2. Bouyer, P., Roux, S.L., Oualhadj, Y., Randour, M., Vandenhove, P.: Games where you can play optimally with arena-independent finite memory. In: Proceedings of the 31st International Conference on Concurrency Theory (CONCUR 2020), LIPIcs, vol. 171, pp. 24:1–24:22 (2020). Schloss Dagstuhl - Leibniz-Zentrum für Informatik

Computing a Fixed Point of a Monotone Function, and Some Applications

Kousha Etessami

School of Informatics, The University of Edinburgh, UK
kousha@inf.ed.ac.uk

Abstract. The task of computing a fixed point of a monotone function arises in a variety of applications.

In this talk I shall describe some recent work in which we have studied the computational complexity of computing a (any) fixed point of a given monotone function that maps a finite d-dimensional grid lattice with sides of length $N = 2^n$ to itself, where the monotone function is presented succinctly via a boolean circuit with $d \cdot n$ input gates and $d \cdot n$ output gates. The underlying ordering, $\leq$, of this lattice is the standard coordinate-wise partial order on d-dimensional vectors in $[N]^d$. By Tarski's theorem, a function $f : [N]^d \to [N]^d$ always has a fixed point if it is monotone, or else it has a pair $x, y \in [N]^d$ that witness non-monotonicity, meaning where $x \leq y$ but $f(x) \not\leq f(y)$. We refer to the corresponding total search problem of either finding a fixed point or finding a witness pair for non-monotonicity, given such a succinctly presented function $f : [N]^d \to [N]^d$, as the `Tarski` problem.

It turns out that `Tarski` subsumes a number of important problems, including some prominent equilibrium computation problems. In particular, we showed that computing the value of Condon's turn-based simple stochastic (reachability) games, as well as the more general problem of computing, within the given desired accuracy $\epsilon > 0$, the value of Shapley's original stochastic games is reducible to `Tarski`. We showed that `Tarski` is contained in both the total search complexity classes **PLS** and **PPAD**. Many questions remain open. I will discuss some of them.

(This talk describes joint work with C. Papadimitriou, A. Rubinstein, and M. Yannakakis, that appeared in the ITCS 2020 proceedings.)

Games with Full, Longitudinal, and Transverse Observability

Orna Kupferman

School of Computer Science and Engineering, The Hebrew University, Israel

Abstract. Design and control of multi-agent systems correspond to the synthesis of winning strategies in games that model the interaction between the agents. In games with *full observability*, the strategies of players depend on the full history of the play. In games with partial observability, strategies depend only on observable components of the history. We survey two approaches to partial observability in two-player turn-based games with behavioral winning conditions. The first is the traditional *longitudinal observability*, where in all vertices, the players observe the assignment only to an observable subset of the atomic propositions. The second is the recently studied *transverse observability*, where players observe the assignment to all the atomic propositions, but only in vertices they own.

O. Kupferman—Supported in part by the Israel Science Foundation, grant No. 2357/19.

Word Equations: From Theory to Practice

Dirk Nowotka

Kiel University, Germany

The existential theory of word equations has been of considerable interest for quite some time. Popularized during the 60's in an attempt to solve Hilbert's 10th problem, it was only shown in 1977 that the theory is decidable by Makanin's seminal work. Since then a large body of work has emerged about solvability, compactness, and complexity questions of word equations. Until today, however, the theory is not fully understood. For example, questions about the membership of solving word equations in NP or the decidability of the existential theory of word equations with length predicate are still open. In recent years, however, the theory of word equations has also gained very practical interest. So called string solvers are becoming more and more interesting in software analysis and access control systems of cloud computing services.

We are going to survey the state of the art of the theory of word equations and sketch its role in practical applications in this talk.

Contents

Invited Papers

On Decision Problems for Substitutions
in Symbolic Dynamics

Valérie Berthé$^{(\boxtimes)}$ (iD)

Université de Paris, IRIF, CNRS, 75013 Paris, France
berthe@irif.fr

Abstract. In this survey, we discuss decidability issues for symbolic dynamical systems generated by substitutions. Symbolic dynamical systems are discrete dynamical systems made of infinite sequences of symbols, with the shift acting on them. Substitutions are simple rules that replace letters by string of letters and allow the generation of infinite words. We focus here on symbolic dynamical systems that are generated by infinite compositions of substitutions, allowing to go beyond the case of the iteration of a single substitution. This is the so-called S-adic framework. Motivated by decidability and ergodic questions, we focus on questions dealing with the convergence of products of nonnegative matrices and associated Lyapunov exponents.

Keywords: Substitutions · Symbolic dynamics · Decidability ·
Lyapunov exponents · Primitive matrix · Perron–Frobenius theorem

1 Introduction

Discrete Dynamical Systems. Dynamical systems describe the evolution of systems over time: a dynamical system is endowed with an evolution rule that describes the time dependence of the state of the elements of the system, allowing to bring out a global average behaviour. One can thus model the evolution of a system whose components interact in a simple way. In physics, this is for instance a set of particles whose state obeys differential equations involving time derivatives, with past determining the future. Modeling using dynamical systems has largely proved its relevance in physics and engineering, and also for numerous phenomena from the digital world, in deep connection with algorithmics: dynamical systems can model the execution of an algorithm, as well as a loop in a program.

In a digital framework, discrete systems arise naturally. More precisely, a discrete (time) dynamical system is defined as the action of a map T acting on a phase space X (usually assumed to be compact), where the rule T governs the discrete time evolution of states in X. Note that the iterative nature of dynamical

This work was supported by the Agence Nationale de la Recherche through the project Codys (ANR-18-CE40-0007).

S. Schmitz and I. Potapov (Eds.): RP 2020, LNCS 12448, pp. 3–19, 2020.
https://doi.org/10.1007/978-3-030-61739-4_1

systems is particularly well adapted to model executions of algorithms. One then studies the evolution of the system in discrete time steps: at time n, we consider the nth iterate T^n. The evolution of the system when starting with the initial condition $x \in X$ is described by the *orbit* or *the trajectory* $(X, Tx, T^2, \cdots, T^n)$ of x. Discrete dynamical systems can be of a geometric nature (e.g., $X = [0, 1]$), or of a symbolic nature (e.g., $X = \{0, 1\}^{\mathbb{N}}$). Consider for instance cellular automata or, here, symbolic dynamical systems, with discrete dynamical systems made of infinite sequences of symbols.

Reachability vs. Statistical Properties of Orbits. Understanding the behavior of orbits allows the global understanding of a dynamical system (X, T). Two types of questions arise naturally in this context. Reachability problems deal with the question of knowing whether an orbit will enter a given subregion Z of X or even reach a given point, whereas the understanding of the long-term behavior allows one to answer the following: will a trajectory visit infinitely often Z (recurrence) and how long will it stay in the subregion Z?

This second type of questions is handled through the use of ergodic theory that describes the long-term statistical behaviour of dynamical systems. An ergodic system is such that the time spent by a trajectory in some region is proportional to the volume of this region; it has the same behavior averaged over time as averaged over all the space. (For a more precise statement, see Sect. 2.)

For linear dynamical systems given by the action of a square matrix with rational entries, reachability problems in the framework of the orbit problem are known to lead to famous number-theoretic undecidable problems, such as the Skolem problem (does a trajectory hit a hyperplane?), the Kannan-Lipton problem [27] (does it hit a point?) and its variants in terms of positivity and ultimate positivity. The synthesis of invariants providing certificates of non-reachability (i.e., invariant sets under the action of T that contain a point x and not y) has opened the way to a large set of applications in verification, control theory, program analysis, etc. See e.g. [22] for the synthesis of semialgebraic invariants.

Substitutions and Positiveness. We focus in the present survey on decision questions that fall within the scope of the ergodic study of the long-term behaviour of orbits. They will conduct us to questions that are related to the existence of occurrences of positive products of matrices.

Dynamically, our main object of study here are substitutions and the words and symbolic dynamical systems they generate. Symbolic dynamical systems are defined on sets of symbols and words with the shift acting on them: the shift is the operation that deletes the first letter of an infinite word. A substitution is a rule, either combinatorial or geometric, that replaces a letter by a string of letters on a finite alphabet, or a tile by a geometric pattern. Iterating substitutions enables the generation of hierarchically ordered structures (infinite words, subshifts, point sets, tilings) that display strong self-similarity properties; one of the best known examples is the Penrose tiling. Substitutions are generalized as

S-adic dynamical systems via a nonstationary (i.e., time inhomogeneous) setting which consists in iterating sequences of substitutions, and not only a single one [8].

For the sake of clarity, we focus here on substitutions acting on words which are more elementary in nature. These simple algorithmic constructions thus produce infinite words and symbolic dynamical systems whose study involves combinatorics on words, ergodic theory, spectral theory, geometry of tilings, Diophantine approximation, number theory, aperiodic order, harmonic analysis, and so on. There exist analogue notions of substitutions and S-adic systems defined on tilings and point sets, and acting as inflation/subdivision rules; see e.g. [34].

Simple refers here to the combinatorial notion of factor complexity of an infinite word with values in a finite alphabet, which counts the number of factors (i.e., strings of consecutive letters that occur in this infinite word) of a given length. This gives an indication of the degree of randomness of this infinite word. The S-adic systems cover a wide class of symbolic dynamical systems with at most linear factor complexity. They were in fact introduced as models for such systems; this is the so-called S-adic conjecture [8,35].

A substitution can efficiently be understood through its incidence matrix (this is somehow the analogue of the incidence matrix of a graph or an automaton). This linear viewpoint enables to exploit a very fruitful dictionary between actions of linear maps and symbolic dynamics, through the matrix/substitution relation produced by the fact that a matrix is the abelianized linear version of a substitution. In particular when this matrix turns out to be primitive, that is, when it admits a power with positive entries, this yields for substitutive subshifts strong ergodic and combinatorial properties, as well as particularly convenient tools for decision problems. Indeed, substitutions are particular cases of free group morphisms (they are free monoid morphisms), with the main simplification being that we have no problem of cancellations. In terms of associated matrices, this gives matrices with nonnegative entries and Perron–Frobenius theory enters into play. However, for the study of S-adic expansions, we do not consider powers of a single matrix but infinite products of nonnegative matrices, and there is no clear analogue of Perron-Frobenius theorem for infinite products of matrices. One relevant ergodic strategy is to go through the theory of Lyapunov exponents at the cost that results are given in a metric way, almost everywhere (see as an illustration Theorem 1). A crucial property in this setting is the existence of occurrences of positive blocks of matrices.

If decision problems for substitutive shifts are well investigated (see Sect. 3.1), in particular thanks to the notion of primitivity, the situation is by nature less effective in the S-adic setting. This survey suggests ways to extend in terms of decidability the substitutive case to the S-adic case (see Sect. 3.2) and asks several natural decision questions (see Sect. 3.3). We also allude to connections with continued fractions, as a source of related questions in Sect. 3.4.

From Symbolic to Arithmetic Dynamics. Let us end this introduction by presenting some elements of motivation for the study of S-adic systems. Symbolic dynamical systems come as codings of trajectories of points in a dynamical system according to a given partition and they also occur in a natural way in arithmetics for instance for the representation of numbers. Symbolic dynamics originates in the work of J. Hadamard in 1898, through the study of geodesics on surfaces of negative curvature. Since its inception, symbolic dynamics has gone hand in hand with substitutions which originated in papers of A. Thue from 1906 and 1912, in particular with the study of the Thue-Morse word. Symbolic dynamics and Sturmian words were developed by Morse and Hedlund in the 40's [32]. Substitutions then turned out to yield unexpected prominent outcomes in the study of quasicrystals and tilings in the framework of aperiodic order. Aperiodic order refers to the mathematical formalization of quasicrystals, initiated by Y. Meyer. Since their discovery in 1982 (for which Nobel prize was awarded to D. Shechtman in 2011), substitutions and aperiodic tilings have proved to be at the heart of their study [5].

One motivation for developing the S-adic formalism comes from the study of algorithms of an arithmetic nature related to specific numeration systems and their applications, running from number theory to cryptography or computer arithmetic. Indeed, in many examples of digital representations, the digits of expansions are produced step by step by the iteration of a transformation and their ergodic study yields information on their digit distribution. This is for instance the case of decimal expansions, beta-numeration, or continued fractions (see e.g. [9, Chapter 2]). As an illustration, consider the dynamical system producing the q-ary expansions of positive real numbers defined as $([0, 1], T_q)$, with $T_q \colon [0, 1] \to [0, 1]$, $x \mapsto qx - \lfloor qx \rfloor = qx \pmod 1$. Indeed, if $x = \sum_{i \geq 1} a_i q^{-i}$, then $a_i = \lfloor qT_q^{i-1}(x) \rfloor$, for $i \geq 1$. When the base q is replaced by some more complicated base β (consider e.g. an algebraic base like the golden ratio $\beta = \frac{1+\sqrt{5}}{2}$), this allows one to expand real numbers in $[0, 1]$ under the form of (possibly infinite) sums of negative powers of β of the form $\sum_{i \geq 1} \varepsilon_i \beta^{-i}$. This is called the beta-numeration.

Algorithms of an arithmetic nature can often be decomposed as a succession of algorithmic steps, which in turn can be modeled by dynamical systems. Kronecker translations $x \mapsto x + \alpha$ modulo 1 provide analogues of the addition, whereas positive entropy systems such as beta-numeration $x \mapsto \beta x$ modulo 1 mirror multiplication. This thus requires a specific nonstationary modeling which consists in working with iterated sequences of transformations that are drawn according to a further dynamical system. More precisely, we do not iterate always the same map T, but the map T can be changed with respect to time: one thus iterates a sequence T_{i_n} of maps acting on phase spaces X_{i_n}. One important feature is the order composition $T_{i_1} \circ \cdots \circ T_{i_n}$. Such a time-inhomogeneous formalism is inspired by the well-studied setting of random dynamics (including random Markov chains and random products of matrices). Symbolic codings of such transformations yields the S-adic formalism.

This survey is organized as follows. Definitions and terminology are gathered in Sect. 2. Specific questions raised by decision issues are then discussed in Sect. 3.

2 General Definitions for Substitutive Dynamical Systems

Ergodic Theory. Ergodic theory studies the long-term average behavior of dynamical systems. One of its main tools is Birkhoff's ergodic theorem which asserts the existence of a time average along each trajectory, provided that (X, T) is endowed with a T-invariant and ergodic measure μ. More precisely, a *measure-theoretic dynamical system* is a dynamical system endowed with an invariant measure $(X, T, \mu, \mathcal{B})$, where μ is a probability measure defined on the σ-algebra $\mathcal{B}$, and $T : X \to X$ is a measurable map which preserves the measure μ, that is, $\mu(T^{-1}(B)) = \mu(B)$ for all $B \in \mathcal{B}$. The measure μ is said to be T-*invariant*. It is said *ergodic* if for every set $B \in \mathcal{B}$, $T^{-1}(B) = B$ has either zero or full measure.

Birkhoff's ergodic theorem states that the time average, that is, $\frac{1}{n} \sum_{k=0}^{n-1} f \circ T^k(x)$ for some observable function f, is the same for almost all initial points x, and is equal to the space average, i.e., $\int_X f \, d\mu$, almost everywhere. The terms almost all and almost everywhere refer to a set of points of X of full measure μ. The ergodic theorem can thus be stated as

$$\forall f \in L^1(X, \mathbb{R}) , \quad \frac{1}{n} \sum_{k=0}^{n-1} f \circ T^k(x) \xrightarrow[n \to \infty]{} \int_X f \, d\mu \quad \mu - a.e.$$

If one takes for f the characteristic function of some subspace Y of X, one deduces that, for almost all trajectories, the proportion of the time spent in Y is equal to the size of Y divided by the size of X (that is, $\mu(Y)/\mu(X)$). As a special case of the ergodic theorem, consider equidistribution theory for sequences defined on the unit interval.

Symbolic Dynamics. Symbolic dynamics offers the advantage of working with coded trajectories, together with combinatorial and topological methods. Consider a finite set $\mathcal{A}$, and let the shift map S act on the set $\mathcal{A}^{\mathbb{N}}$ of infinite words with values in the alphabet $\mathcal{A}$ as $S((u_n)_{n \in \mathbb{N}}) = (u_{n+1})_{n \in \mathbb{N}}$. (Note that all the notions and results here hold also for biinfinite words in $\mathcal{A}^{\mathbb{Z}}$.) Here, the set $\mathcal{A}^{\mathbb{N}}$ is equipped with the product topology of the discrete topology on each copy of $\mathcal{A}$. Thus, this set is a compact space. This topology is the topology defined by the following distance: for $u \neq v \in \mathcal{A}^{\mathbb{N}}$, $d(u, v) = 2^{-\min\{n \in \mathbb{N}; \ u_n \neq v_n\}}$. Thus, two infinite words are close to each other if their first terms coincide.

A *subshift* is a closed shift invariant subset of some $\mathcal{A}^{\mathbb{N}}$ for $\mathcal{A}$ finite alphabet. A *factor* of an infinite word u is a string of consecutive letters that occurs in u. The *factor complexity* of an infinite word is the function that counts the number of factors of a given length that occur in it. The set of factors $\mathcal{L}_u$ of an infinite word u is called its *language*. This definition extends to the language of a subshift (X, S): this is the set of factors of infinite words in X. One important feature of a subshift is that it is defined by its language, since it is closed and shift-invariant. Subshifts of *finite type* are then defined as the subshifts whose set of words (in their language) is finite. *Sofic subshifts* are images of subshifts of finite type under a factor map, where a factor map $\pi : X \to Y$ between two subshifts X and Y is a continuous, surjective map such that $\pi \circ S = S \circ \pi$. A subshift (X, S) is *minimal* if X has no nontrivial closed shift-invariant subset; then, all the infinite words in X have the same language.

The *frequency* of a letter i in an infinite word u is defined as the limit when n tends towards infinity, if it exists, of the number of occurrences of i in $u_0 u_1 \cdots u_{n-1}$ divided by n. Frequencies of factors are defined analogously. Let (X, S) be a minimal subshift. For a given (finite) word w of the language of X, the cylinder $[w]$ is the set of infinite words in X that have w as a prefix, i.e., $[w] = \{v \in X; v_0 \ldots v_{n-1} = w\}$. If μ is an ergodic measure on u, then we deduce from Birkhoff's ergodic theorem that, for μ-almost every infinite word in X, any w has frequency $\mu([w])$. If the shift (X, S) is uniquely ergodic (i.e., there exists a unique shift-invariant probability measure on X), then the unique invariant measure on X is ergodic and the convergence in Birkhoff's ergodic theorem holds uniformly for every infinite word in X. For more details on invariant measures and ergodicity, we refer to [36] and [9, Chap. 7].

Primitive Substitutions. A *substitution* σ is an application from an alphabet $\mathcal{A}$ into the set of nonempty finite words on $\mathcal{A}$; it extends to a morphism of the free monoid $\mathcal{A}^*$ by concatenation, that is, $\sigma(ww') = \sigma(w)\sigma(w')$. It also extends in a natural way to a map defined over $\mathcal{A}^{\mathbb{N}}$. The substitutive symbolic dynamical system (X_σ, S) generated by σ is then defined as the set of infinite words w such that there exists a letter a and a nonnnegative integer n such that w is a factor of $\sigma^n(a)$.

Substitutions are very efficient tools for producing infinite words. As an example, consider the substitution σ on $\mathcal{A} = \{a, b\}$ defined by $\sigma(a) = ab$ and $\sigma(b) = a$. Then, the sequence of finite words $(\sigma^n(a))_n$ starts with $\sigma^0(a) = a$, $\sigma^1(a) = ab$, $\sigma^2(a) = aba$, $\sigma^3(a) = abaababa, \ldots$ Each $\sigma^n(a)$ is a prefix of $\sigma^{n+1}(a)$, and the limit word in $\mathcal{A}^{\mathbb{N}}$ is

$$abaababaabaababaabaababaabaababaabaababaabaababaabaababaabaababaabaab\ldots$$

The above limit word is called the *Fibonacci word* (for more on the Fibonacci word, see e.g. [31, 35]).

For $i \in \mathcal{A}$ and for $w \in \mathcal{A}^*$, let $|w|_i$ stand for the number of occurrences of the letter i in the word w. Let d stand for the cardinality of $\mathcal{A}$. The *incidence matrix* M_σ of the substitution σ is the square matrix with entries $|\sigma(j)|_i$. This is a commutative linear version of the substitution σ and it offers the advantage to bring all the strength of Perron-Frobenius theorem for nonnegative matrices (see e.g. [39]). A substitution is said *primitive* if there exists a power of its incidence matrix whose entries are all positive. According to Perron–Frobenius theorem, if a substitution is primitive, then its incidence matrix admits a dominant eigenvalue (it dominates strictly in modulus the other eigenvalues) that is (strictly) positive. It is called its *Perron–Frobenius eigenvalue*.

The dynamical system (X_σ, S) associated with a primitive substitution σ can be endowed with a natural shift-invariant Borel probability measure μ defined by its values on the cylinders. The measure of the cylinder $[w]$ is defined as the frequency of the finite word w in any infinite word of X_σ, which is proved to exist, with a proof based again on Perron–Frobenius theorem [36].

Primitive substitutions have numerous interesting properties: the subshift (X_σ, S) is minimal, linearly recurrent, uniquely ergodic and any of its elements has at most linear factor complexity. (An infinite word u is said to be *linearly recurrent* if there exists a constant C such that $R(n) \le Cn$, for all n.) For more details, see [36].

S-adic Words. An S-adic dynamical system is defined in terms of a sequence of substitutions. Let S be a set of substitutions. Let $s = (\sigma_n)_{n \in \mathbb{N}} \in S^{\mathbb{N}}$, with $\sigma_n : \mathcal{A}_{n+1}^* \to \mathcal{A}_n^*$, be a sequence of substitutions, and let $(a_n)_{n \in \mathbb{N}}$ be a sequence of letters with $a_n \in \mathcal{A}_n$ for all n. We say that the infinite word $u \in \mathcal{A}^{\mathbb{N}}$ admits $((\sigma_n, a_n))_n$ as an *S-adic representation* if

$$u = \lim_{n \to \infty} \sigma_0 \sigma_1 \cdots \sigma_{n-1}(a_n).$$

The sequence s is called the *directive sequence* and the sequences of letters $(a_n)_n$ will only play a minor role compared to the directive sequence. If the set S is finite, it makes no difference to consider a constant alphabet (i.e., $\mathcal{A}_n^* = \mathcal{A}^*$ for all n and for all substitution σ in S). Note that the S-adic dynamical system of a periodic directive sequence $(\sigma_0, \cdots, \sigma_{n-1})^\infty$ is equal to the substitutive dynamical system generated by the substitution $\sigma_0 \circ \cdots \circ \sigma_{n-1}$.

To be "S-adic" is not an intrinsic property of an infinite word, but a way to construct it and an infinite word admits many possible S-adic representations [8]. But some S-adic representations might be useful to get information about the word. One thus adds the following requirement. An S-adic representation defined by the directive sequence $(\sigma_n)_{n \in \mathbb{N}}$ is *everywhere growing* if for any sequence of letters $(a_n)_n$, one has

$$\lim_{n \to +\infty} |\sigma_{[0,n)}(a_n)| = +\infty,$$

with $\sigma_{[0,n)} := \sigma_0 \circ \sigma_{k+1} \ldots \circ \sigma_{n-1}$.

Given an everywhere growing directive sequence $(\sigma_n)_{n \in \mathbb{N}}$ of substitutions that are all defined over the same finite alphabet $\mathcal{A}$, the *subshift* associated with $(\sigma_n)_n$ is defined as the set of infinite words whose set of factors is included in some $\sigma_{[0,n)}(i)$, for some $i \in \mathcal{A}$.

As a prominent example, consider Sturmian words for which the Fibonacci word is a particular case. The substitutions τ_a and τ_b are defined over the alphabet $\mathcal{A} = \{a, b\}$ by $\tau_a : a \mapsto a, b \mapsto ab$ and $\tau_b : a \mapsto ba, b \mapsto b$. Let $(i_n) \in \{a, b\}^{\mathbb{N}}$. The following limits

$$u = \lim_{n \to \infty} \tau_{i_0} \tau_{i_1} \cdots \tau_{i_{n-1}}(a) = \lim_{n \to \infty} \tau_{i_0} \tau_{i_1} \cdots \tau_{i_{n-1}}(b) \tag{1}$$

exist and coincide whenever the directive sequence $(i_n)_n$ is not ultimately constant (it is easily shown that the shortest of the two images by $\tau_{i_0} \tau_{i_1} \ldots \tau_{i_{n-1}}$ is a prefix of the other). The infinite words thus produced belong to the class of Sturmian words, and a *Sturmian word* is an infinite word whose set of factors coincides with the set of factors of a sequence u of the form (1), with the sequence $(i_n)_{n \geq 0}$ being not ultimately constant.

We have seen that the notion of primitivity plays an important role for substitutions. There are two ways of extending this notion in the S-adic setting. An S-adic expansion with directive sequence $(\sigma_n)_n$ is said *weakly primitive* if, for each n, there exists r such that the substitution $\sigma_n \cdots \sigma_{n+r}$ is positive. It is said *strongly primitive* if the set of substitutions $\{\sigma_n\}$ is finite, and if there exists r such that the substitution $\sigma_n \cdots \sigma_{n+r}$ is positive, for each n.

The following statement from [16] illustrates the role of primitivity in the S-adic context: if a directive sequence is weakly primitive, then the associated shift is minimal. If it is strongly primitive, the associated shift is minimal, uniquely ergodic, and it has at most linear factor complexity. With an extra condition of properness one even obtains the following characterization of linear recurrence [16]: a subshift (X, S) is linearly recurrent if and only if it is a strongly primitive and proper S-adic subshift. (A substitution over $\mathcal{A}$ is said *proper* if there exist two letters $b, e \in \mathcal{A}$ such that for all $a \in \mathcal{A}$, $\sigma(a)$ begins with b and ends with e. An S-adic system is said to be *proper* if the substitutions in S are proper.) An essential ingredient in the proofs of these results is the uniform growth of the matrices $M_{(0,n)}$ as for substitutive systems.

Discrepancy and Balancedness Properties. Ergodic deviations control the convergence of ergodic sums and measure the difference between $1/n \sum_{k=0}^{n-1} f \circ T^k(x)$ and the expected value $\int f d\mu$. Of course, ergodic deviations depend on the nature of the dynamical system and on the regularity of f. If f is the indicator function of some given subset of X, this corresponds to the classical notion of discrepancy in equidistribution theory. Discrepancy measures the difference between the actual number of points in a subset and the expected number of points. This makes particularly sense in a metric number-theoretic framework for Kronecker sequences $(n\alpha)_n \mod 1$. Low discrepancy sequences are widely used in the Monte Carlo method and dynamical systems may be applied to generate higher-dimensional low-discrepancy sequences.

In symbolic dynamics, ergodic deviations measure the convergence toward frequencies, via symbolic discrepancy. Counting frequencies of words in a given symbolic dynamical system (or in a tiling space) is among the most fundamental questions of the field. Recurrence (repetitiveness for tilings and point-sets) is a closely related fundamental notion of order which describes how often a finite given configuration occurs (see e.g. [9, Chap. 7]).

The symbolic version of discrepancy is defined as follows [1]. Let $u \in \mathcal{A}^{\mathbb{N}}$ be an infinite word and assume that each letter i has frequency f_i in u. The *letter discrepancy* of u is $\Delta_n(u) = \sup_{i \in \mathcal{A}} ||u_0 u_1 \dots u_{n-1}|_i - n f_i|$. It can also be defined by general factors (not only letters).

Symbolic letter discrepancy is closely related to balancedness, which is a measure of disorder which counts the difference between the numbers of occurrences of a given word in factors of the same length. An infinite word $u \in \mathcal{A}^{\mathbb{N}}$ is said to be C-*balanced* if for any pair v, w of factors of the same length of u, and for any letter $i \in \mathcal{A}$, one has $||v|_i - |w|_i| \leq C$. It is said *balanced* if there exists $C > 0$ such that it is C-balanced. Then, if the letters have frequencies in u, u is balanced if and only if its letter discrepancy is finite (uniformly in n).

As examples of balanced words, Sturmian words are known to be 1-balanced; they even are exactly the 1-balanced infinite words that are not eventually periodic [31]. Balance for substitutions can also be studied via the incidence matrix and a characterization of balanced words generated by primitive substitutions is given in [1]. Indeed, let σ be a primitive substitution and λ be its Perron-Frobenius eigenvalue; if its second eigenvalue is smaller than 1, then the letter discrepancy is finite. These results can be extended from letters to words.

Generalized Perron-Frobenius Eigenvectors. Given a directive sequence $(\sigma_n)_n$ that is everywhere growing, the cone $\bigcap_n M_{[0,n)} \mathbb{R}^d_+$ determined by the incidence matrices of the substitutions σ_n is intimately related to letter frequencies in the corresponding S-adic shift: it is the convex hull of the set of half lines $\mathbb{R}_+ \mathbf{f}$ generated by the vector $\mathbf{f}$ whose components are the frequencies of letters of infinite words in the associated S-adic shift.

In the primitive substitutive case, letter frequencies are given by the Perron-Frobenius eigenvector [36]. For a primitive matrix M, the cones $M^n \mathbb{R}^d_+$ nest down to a single line directed by this eigenvector at an exponential convergence speed, according to Perron–Frobenius theorem (see e.g. [39]).

Concerning primitivity, the situation is more contrasted for S-adic systems. For instance, weak primitivity is known not to imply unique ergodicity [28]. Let $M_{[0,n)} := M_0 M_1 \cdots M_{n-1}$. We now state a sufficient condition for the sequence of cones $M_{[0,n)} \mathbb{R}^d_+$ to nest down to a single strictly positive direction as n tends to infinity (provided that the square matrices M_n have all non-negative entries); in other words, the columns of the product $M_{[0,n)}$ tend to be proportional. The following convenient sufficient condition [23] is thus a (weak) analogue of Perron-Frobenius theorem. It is stated in terms of infinite occurrences of a positive block of matrices. Let $(M_n)_n$ be a sequence of non-negative integer matrices of size d. Assume that there exist a strictly positive matrix B and indices $j_1 < k_1 \leq j_2 < k_2 \leq \cdots$ such that $B = M_{j_1} \cdots M_{k_1-1} = M_{j_2} \cdots M_{k_2-1} = \cdots$. Then, there exists a vector $\mathbf{f} \in \mathbb{R}^d_+$ with positive entries such that the sequence of cones $M_{[0,n)} \mathbb{R}^d_+$ nests down to the direction carried by $\mathbf{f}$ as n tends to infinity, i.e., $\bigcap_{n \in \mathbb{N}} M_{[0,n)} \mathbb{R}^d_+ = \mathbb{R}_+ \mathbf{f}$. It is widely used in symbolic dynamics [12, 44].

The proof of this theorem relies on classical methods for non-negative matrices, namely Birkhoff contraction coefficient estimates and projective Hilbert metric. The vector $\mathbf{f}$ is a generalized right eigenvector of $\boldsymbol{\sigma}$. In particular, the letter frequency vector is a generalized right eigenvector for an infinite S-adic word generated by a sequence of substitutions whose incidence matrices satisfy the above conditions. Note that there is no way to define a generalized left eigenvector as the sequence of rows vary significantly in the sequence of matrices $(M_{[0,n)})_n$, in contrast to the columns. In the case of a single matrix, one simply takes the transpose, with the cones nesting down for rows as well as for columns.

Lyapunov Exponents. We can go beyond the previous sufficient condition (yielding the existence of a generalized right eigenvector), at the cost of working in average. Lyapunov exponents then replace (logarithms) of Perron–Frobenius eigenvalues. They describe the asymptotic behaviour of the singular values of large products of random matrices, under the ergodic hypothesis.

Let $\mathcal{S}$ be a finite set of substitutions with invertible incidence matrices, and let (D, S, μ) with $D \subset \mathcal{S}^{\mathbb{N}}$ be an ergodic subshift equipped with a probability measure μ. It can be for instance a shift of finite type or a sofic shift. Here S stands also for the shift acting on D. We thus have a second dynamical system governing the substitutions to be iterated. Given an infinite sequence of substitutions $\sigma = (\sigma_n)_n \in D$, we define $M_{[0,n]}(\sigma) := M_0 M_1 \cdots M_{n-1}$, where M_i is the incidence matrix of the substitution σ_i. The Lyapunov exponents of the products $M_{[0,n]}(\sigma)$ with respect to the ergodic probability measure μ provide the exponential growth of eigenvalues of the matrices $M_{[0,n]}(\sigma)$ along a μ-generic sequence σ. The existence of Lyapunov exponents generalizes Birkhoff's ergodic theorem in a non-commutative setting. The fact that $\mathcal{S}$ is a finite set of substitutions with invertible incidence matrices ensures log-integrability, that is, $\int_D \log \max(\|M_1(\sigma)\|, \|M_1(\gamma)^{-1}\|) d\mu(\sigma) < \infty$. By ergodicity of μ, the first Lyapunov exponent of (D, S, ν) is the μ-almost everywhere[1] limit

$$\theta_1^{\mu} = \lim_{n \to \infty} \frac{\log \|M_{[0,n)}(\sigma)\|}{n}.$$

The other Lyapunov exponents $\theta_2^{\mu} \geq \theta_3^{\mu} \ldots \geq \theta_d^{\mu}$ are defined recursively by the following μ-almost everywhere limits, for $k = 1, \ldots, d$:

$$\theta_1^{\mu} + \theta_2^{\mu} + \cdots + \theta_k^{\mu} = \lim_{n \to \infty} \frac{1}{n} \int_D \log \| \wedge^k M_{[0,n)} \| \, d\nu$$

where $\wedge^k$ stands for the k-th exterior product (the k-fold wedge product). We are mostly interested by the two first Lyapunov exponents θ_1^{μ} and θ_2^{μ} that govern the convergence of column vectors for the products $M_{[0,n)}$.

Lyapunov exponents enable to state convergence results (see Theorem 1 below). By following the vocabulary of Markov chains [39], or of continued fractions [38], it is natural to consider the following definitions for convergence for the columns of the products of nonnegative matrices $M_{[0,n)}$. Assume the existence of a generalized right eigenvector $\mathbf{f}$. We work on the alphabet $\mathcal{A} = \{1, \ldots, d\}$. Let $\mathbf{f}_i^{(n)}$ stand for the column vectors of $M_{[0,n)}$. The column vectors $\mathbf{f}_i^{(n)}$, $1 \leq i \leq d$, produce d sequences of *rational convergents* $(\mathbf{f}_i^{(n)}/\|\mathbf{f}_i^{(n)}\|_1)_{n \in \mathbb{N}}$ that converge to $\mathbf{f}$.

More precisely,

- the convergence is said to be *weak* if $\lim_{n \to \infty} \mathbf{f}_i^{(n)}/\|\mathbf{f}_i^{(n)}\|_1 = \mathbf{f}$ holds for all $i \in \{1, \ldots, d\}$;
- it is said to be *strong* if $\lim_{n \to \infty} \|\mathbf{f}_i^{(n)} - \|\mathbf{f}_i^{(n)}\|_1 \mathbf{f}\| = 0$ holds for all $i \in \{1, \ldots, d\}$;

---

[1] Here μ-almost everywhere refers to directive sequences of substitutions chosen in D with respect to the measure μ.

– it is said *exponential* if there are positive constants $\kappa, \delta \in \mathbb{R}$ such that $\|\mathbf{f}_i^{(n)} - \|\mathbf{f}_i^{(n)}\|_1 \mathbf{x}\| < \kappa e^{-\delta n}$ holds for all $i \in \{1, \ldots, d\}$ and all $n \in \mathbb{N}$.

Weak convergence means that the angle between the column vectors and $\mathbf{f}$ tends to 0 whereas strong convergence means that the distance between the column vectors and $\mathbf{f}$ tends to 0. In the case of a single primitive nonnegative matrix, Perron–Frobenius theorem yields exponential convergence.

Working with Lyapunov exponents provides the following statement [8]. Note that one of its limits, in terms of effectiveness, relies in the fact that it holds almost every where, that is, for a set of full measure with respect to the measure μ. This is often a very delicate task to be able to describe in effective terms such a full measure set.

Theorem 1. *Let $\mathcal{S}$ be a finite set of substitutions with invertible incidence matrices, and let $(D, \mathcal{S}, \mu)$, with $D \subset \mathcal{S}^{\mathbb{N}}$, be an ergodic shift. Assume that there exists a product of substitutions $\sigma_0 \ldots \sigma_{k-1}$ with positive incidence matrix $M_{\sigma_0} \cdots M_{\sigma_{k-1}}$ whose associated cylinder in D has positive measure for μ. For μ-almost every directive sequence of substitutions $\sigma \in D$, the corresponding $\mathcal{S}$-adic system X_σ is minimal and uniquely ergodic and one has weak convergence. Furthermore, $\theta_1^\mu > 0$ and $\theta_1^\mu > \theta_2^\mu$. If $\theta_2^\mu < 0$, then, for μ-almost every S-adic sequence in D, (X_σ, S) has bounded letter discrepancy and the convergence is strong.*

The quantity $1 - \frac{\theta_2^\mu}{\theta_1^\mu}$ can also be found in the context of continued fractions [29] as the uniform approximation exponent for multidimensional continued fractions algorithms such as the Jacobi-Perron algorithm. We will come back to it in Sect. 3.4. The existence of a product of substitutions $\sigma_0 \ldots \sigma_{k-1}$ with positive incidence matrix is crucial here and plays the role of primitivity.

3   From Substitutions to S-adic Systems

Since substitutions are finite in nature, many of their properties are decidable. In particular, we can decide whether a matrix is primitive and we have seen with Sect. 2 the interest of being primitive for substitutions. We first recall decidability properties for primitive substitutions in Sect. 3.1. The next question is to have a suitable decision framework for S-adic shifts. Elements in this direction are given in Sect. 3.2. Specific decision questions are listed in Sect. 3.3. We conclude by developing an analogy with continued fractions in Sect. 3.4.

3.1   Some Decisions Problems for Substitutions

Numerous decidability results exist for fixed points of substitutions and their images by general morphisms (see e.g. [20, Section 10]). For primitive substitutions, various decision problems have been proved recently using the notion of return words and derived sequences. A return word to a factor u of an infinite word is a factor w of this infinite word such that uw admits exactly two occurrences of u, with the second occurrence of u being a suffix of uw. One then can recode infinite words generated by a primitive substitution via return words, obtaining derived sequences (see e.g. [16]). This has allowed the resolution of several long-standing decision problems. Let us quote the decidability of the equality between two substitutive infinite words [17]; see also [9, Chapter 10]. The decidability of the ultimate periodicity of substitutive infinite words is also decidable (see [17] for the primitive case, and [18] for the general case) as well as the uniform recurrence of substitutive words [19]. This problem is closely related to the decidability of the ultimate periodicity of recognizable sets of integers in some abstract numeration systems [6]. One can even decide whether two minimal substitution subshifts are topologically isomorphic and even whether one is a factor of the other [20]. In the particular case of constant-length substitutions (automatic sequences), the connections between first-order logic and automata produce efficient decision procedures (see e.g. [10,15,41]) relying on the equivalence between being p-recognizable and p-definability [14].

More generally, decidability in symbolic dynamics has already a long history since the undecidability of the emptiness problem (the domino problem) for multi-dimensional subshifts of finite type [7,37]; see also [9, Chapter 8]. Since the beginning, substitutions were used to input computation in tilings and they produced the first examples of aperiodic tilings, such as Robinson or Penrose tilings, proving the undecidability of the domino problem. Moreover, computability is a notion that has appeared as a major understanding tool in the study of multi-dimensional subshifts of finite type with the breakthrough characterization of the entropies of multi-dimensional subshifts of finite type as the non-negative right recursively enumerable numbers [25]. Let us mention also the realization of effective subshifts (with factor and projective subaction operation) from higher-dimensional subshifts of finite type [3,21,24]; see also [9, Chapter 9] and [26].

3.2   On Effectiveness in the S-adic Framework

Decision problems for substitutions make particular sense since the data that describe substitutions are finite. However, the description of an S-adic system is not finite, it is based on the infinite directive sequence of substitutions. The first issue which arises in this context is thus to give a meaning to effectiveness. There are several notions that can be considered and that are intimately related [11]: effectiveness of the directive sequence, computability of frequencies/invariant measures, and decidability of the language. We describe them below.

We recall that a subshift X can be defined by providing its language, that is, the set of finite words that occur in infinite words in X. It can be defined equivalently by providing the set of forbidden factors. This leads to the following definitions. A subshift is said to be Π_1-computable or effective if its language is co-recursively enumerable; it is said Σ_1-computable if its language is recursively enumerable; it is said Δ_1-computable or decidable if its language is recursive. A subshift (X, S) is said to have computable frequencies if the frequencies of factors exist and are uniformly computable. A shift-invariant measure is said to be computable if the measure of any cylinder is uniformly computable. A closed subset $D \subset S^{\mathbb{N}}$ is *effectively closed* if the set of (finite) words which do not appear as prefixes of elements of S is recursively enumerable (one enumerates forbidden prefixes). An effectively closed set is not necessarily a subshift.

The following relations between these concepts offers a convenient framework for decision problems for S-adic shifts (see [11]). Let X be a subshift. If X is effective and uniquely ergodic, then its invariant measure is computable and X is decidable. If X is minimal and its frequencies are computable, then its language is recursively enumerable. If X is minimal and effective, then it is decidable. If it is minimal, uniquely ergodic, and defined with respect to a directive sequence $\sigma \in S^{\mathbb{N}}$, then the following conditions are equivalent: there exists a computable directive sequence σ' such that $X_\sigma = X_{\sigma'}$; the unique invariant measure of X_σ is computable; the subshift X_σ is decidable.

3.3 Around Decision Problems for S-adic Systems

The concepts described in Sect. 2 lead to numerous decision questions. For instance, can we decide rational independence for the coordinates of a generalized right eigenvector? Note that sufficient conditions can be given in terms of strong convergence [12]. Concerning classification issues, can the isomorphism between two (effective) S-adic systems be decided (in the spirit of [20] which handles the substitutive case)? Can we decide whether balancedness on letters holds? and on factors? Can we decide recurrence properties such as having linear recurrence? Due to the fact that we lose the self-similarity properties present for substitutive systems, such results require new ideas and do not run along the lines of the substitutive setting.

We have seen the importance of Lyapunov exponents. Their computation also raises specific questions. Hardness is considered in [42] where the largest Lyapunov exponent is proved not to be algorithmically approximable. Practically, ergodic theorems provide efficient ways of estimating numerically Lyapunov exponents by following trajectories and then taking averages over truncated trajectories. Moreover, with ergodic theory and probability come methods issued from thermodynamic formalism, and more particularly transfer operators. Indeed, the theory of transfer operators can be considered as the analogue for invariant densities of dynamical systems of Perron–Frobenius theory for nonnegative matrices. They govern the evolution in time of the mass distribution of points and their action on densities models the action a dynamical system on input distributions. Since they are bounded linear operators, this makes them suitable for computer approximations (via truncations with respect to functional spaces) [30,33]. Note that the study of extremal matrix products and of the joint spectral radius (the largest asymptotic rate of growth that can be obtained by forming long products of matrices) is also particularly relevant in symbolic dynamics (see e.g. [9, Chapter 11]).

Pisot dynamics is a further specific setting that leads to numerous decision problems. A Pisot number is an algebraic integer whose algebraic conjugates (other than itself) are smaller than 1. In the eighties, Pisot substitutions (i.e., primitive substitutions whose Perron–Frobenius factor is a Pisot number) attracted much attention in the context of mathematical quasicrystals. We recall that quasicrystals are solids that are ordered but not periodic, and since their discovery, fruitful mathematical formulations have been proposed for the understanding of how atoms must be arranged in a material for it to diffract like a quasicrystal [5]. Pisot substitutions play a crucial role here since they create a hierarchical structure with a significant amount of long range order. They are conjectured to have pure discrete spectrum, that is, to be isomorphic (in the measure-theoretic sense) to a translation on a compact abelian group [2,9]. In a nutshell, algebraicity plus the Pisot arithmetic condition equals order. Order is expressed here in spectral and dynamical terms as being isomorphic to the simplest dynamical systems, namely group translations. This conjecture remains open since the 80's. For Pisot substitutions, several tools based on graphs are proposed in order to decide pure discrete spectrum [40]. In [12], the Pisot conjecture is extended to the S-adic setting going beyond algebraicity. The Pisot condition is then replaced by the requirement that the second Lyapunov exponent of the dynamical system is negative. Deciding pure discrete spectrum has to be formulated in this context.

Note that the study of Sturmian words and their various extensions is a setting that has confirmed the crucial role played by primitivity. Indeed, the S-adic expansion of Sturmian words is governed by continued fractions that play a renormalization role via the geodesic flow [35]. This has been very successfully extended with the study of interval exchanges in relation with the Teichmüller flow. Finding occurrences of positive matrices in the associated infinite products of matrices is at the heart of their study (see e.g. [4,43,44]).

3.4   Continued Fractions and S-adic Expansions

Decision questions in the S-adic setting are nourished by the dictionary that exists between continued fractions and S-adic expansions. Indeed, one important principle governing the S-adic approach relies in the translation of a continued fraction expansion in symbolic terms via the matrix/substitution correspondence between a substitution and its incidence matrix. A continued fraction algorithm in dimension d generates products of nonnegative square matrices $(M_n)_n$ of size d, with the expanded vector belonging to the cone $\cap_n M_1 \cdots M_n \mathbb{R}_+^d$. A continued fraction algorithm then provides an S-adic subshift.

More generally, continued fraction expansions provide increasingly good rational approximations of real numbers. A continued fraction is expected to yield simultaneous better and better rational approximations with the same denominator for d-uples $(\alpha_1, \cdots, \alpha_d)$ in $[0,1]^d$, in an effective way and with a good approximation quality. It has to produce a sequence of positive integers $(q_n)_n$ such that the distance to the nearest integer $|||q_n(\alpha_1, \cdots, \alpha_d)|||$ converges exponentially fast to 0 with respect to q_n, and ideally in $q_n^{-\frac{1}{d}}$, with respect to Dirichlet's theorem.

There is no canonical extension of regular continued fractions to higher dimensions (the monoid $SL(3,\mathbb{N})$ is not free, contrarily to $SL(2,\mathbb{N})$), and the zoology of existing types of algorithms is particularly rich. Indeed, regular continued fractions rely on Euclid algorithm: starting with two numbers, one subtracts the smallest from the largest. If we start with at least three numbers, it is not clear to decide which operation has to be performed on these numbers in order to get something analogous to Euclid algorithm, hence the diversity and multiplicity of existing generalizations. Famous examples are the Jacobi-Perron, Brun or Selmer algorithms [38].

Note that in this correspondence, substitutions are associated with matrices in a noncanonical way. Indeed, a matrix can be the incidence matrix of several substitutions. A substitution offers in fact more information than a matrix. Given a continued fraction algorithm, a first step is thus to choose correctly the substitutions associated with matrices. Once a suitable choice of substitutions will be provided, a continued fraction algorithm will provide an S-adic system, it remains to investigate its combinatorial properties (factor complexity, recurrence and frequencies, symbolic discrepancy).

The main advantage of most classical unimodular continued fractions is that they can be expressed as dynamical systems whose ergodic study has already been well understood [38]. Convergence issues are then to be discussed and statements such as Theorem 1 have to be made effective. However, in higher dimension, continued fraction algorithms seem to present a major drawback concerning the quality of approximation. The convergence is governed by the quantity $1 - \frac{\theta_2}{\theta_1}$ [29], where θ_1 and θ_2 are the two largest Lyapunov exponents of the associated dynamical system) (cf. Theorem 1). It has to be compared with Dirichlet's exponent $1 + 1/d$. However, recent striking numerical experimentations [13] indicate that the second Lyapunov exponent is not even negative for the most classical continued fraction algorithms, such as the Brun, Jacobi-Perron or Selmer algorithms in dimension d with $d \geq 10$, contrarily to what was expected. In other words, strong convergence is lost when increasing the dimension. A first challenge is to confirm these experimental results theoretically. This also raises the need for designing efficient strongly convergent continued fraction algorithms in dimension larger than 2, conducting to S-adic systems, thus providing further decision questions to explore.

References

1. Adamczewski, B.: Symbolic discrepancy and self-similar dynamics. Ann. Inst. Fourier (Grenoble) **54**, 2201–2234 (2004)
2. Akiyama, S., Barge, M., Berthé, V., Lee, J.-Y., Siegel, A.: On the Pisot substitution conjecture. In: Kellendonk, J., Lenz, D., Savinien, J. (eds.) Mathematics of Aperiodic Order. PM, vol. 309, pp. 33–72. Springer, Basel (2015). https://doi.org/10.1007/978-3-0348-0903-0_2
3. Aubrun, N., Sablik, M.: Simulation of effective subshifts by two-dimensional subshifts of finite type. Acta Applicandae Mathematicae **126**, 35–63 (2013)
4. Avila, A., Forni, G.: Weak mixing for interval exchange transformations and translation flows. Ann. Math. **165**, 637–664 (2007)
5. Baake, M., Grimm, U.: Aperiodic order. Volume 1: A Mathematical invitation. Cambridge University Press, Cambridge (2013)

6. Bell, J.P., Charlier, E., Fraenkel, A.S., Rigo, M.: A decision problem for ultimately periodic sets in non-standard numeration systems. Int. J. Algebra Comput. **9**, 809–839 (2009)
7. Berger, R: The undecidability of the domino problem. Mem. Am. Math. Soc. **66** (1966)
8. Berthé, V., Delecroix, V.: Beyond substitutive dynamical systems: S-adic expansions. RIMS Lecture note 'Kokyuroku Bessatu' **B46**, 81–123 (2004)
9. Berthé, V., Rigo, M. (eds.): Combinatorics, Automata and Number Theory. Encyclopedia of Mathematics and its Applications, vol. 135. Cambridge University Press, Cambridge (2010)
10. Berthé, V., Rigo, M. (eds.): Combinatorics, Words and Symbolic dynamics. Encyclopedia of Mathematics and its Applications, vol. 159, Cambridge University Press (2016)
11. Berthé, V., Fernique, T., Sablik, M.: Effective S-adic symbolic dynamical systems. In: Beckmann, A., Bienvenu, L., Jonoska, N. (eds.) CiE 2016. LNCS, vol. 9709, pp. 13–23. Springer, Cham (2016). https://doi.org/10.1007/978-3-319-40189-8_2
12. Berthé, V., Steiner, W., Thuswaldner, J.: Geometry, dynamics, and arithmetic of s-adic shifts. Annales de l'Institut Fourier **69**, 1347–1409 (2019)
13. Berthé, V., Steiner, W., Thuswaldner, J.: On the strong convergence of higher-dimensional continuous fractions. Math. Comput. (to appear)
14. Bruyère, V., Hansel, G., Michaux, C., Villemaire, R.: Logic and p-recognizable sets of integers. Bull. Belg. Math. Soc. Simon Stevin **1**(2), 191–238 (1994)
15. Charlier, E., Rampersad, N., Shallit, J.: Enumeration and decidable properties of automatic sequences. Int. J. Found. Comput. Sci. **23**, 1035–1066 (2012)
16. Durand, F.: Linearly recurrent subshifts have a finite number of non-periodic subshift factors. Ergod. Theory Dyn. Syst. **20**, 1061–1078 (2000). Corrigendum and addendum. Ergodic Theory Dynam. Systems, **23** 663–669 (2003)
17. Durand, F.: HD0L ω-equivalence and periodicity problems in the primitive case. Unif. Distrib. Theory **7**, 199–215 (2012)
18. Durand, F.: Decidability of the HD0L ultimate periodicity problem. RAIRO Theor. Inform. Appl. **47**, 201–214 (2013)
19. Durand, F.: Decidability of uniform recurrence of morphic sequences. Int. J. Found. Comput. Sci. **24**, 123–146 (2013)
20. Durand, F., Leroy. J.: Decidability of the isomorphism and the factorization between minimal substitution subshifts. https://arxiv.org/abs/1806.04891
21. Durand, B., Romashchenko, A., Shen, A.: Effective closed subshifts in 1D can be implemented in 2D. In: Blass, A., Dershowitz, N., Reisig, W. (eds.) Fields of Logic and Computation. LNCS, vol. 6300, pp. 208–226. Springer, Heidelberg (2010). https://doi.org/10.1007/978-3-642-15025-8_12
22. Fijalkow, N., Ohlmann, P., Ouaknine, J., Pouly, A., Worrell, J.: Complete semialgebraic invariant synthesis for the Kannan-Lipton orbit problem. Theor. Comput. Sci. **63**, 1027–1048 (2019)
23. Furstenberg, H.: Stationary Processes and Prediction Theory. Annals of Mathematics Studies, vol. 44, Princeton University Press, Princeton (1960)
24. Hochman, M.: On the dynamics and recursive properties of multidimensional symbolic systems. Inventiones Mathematicae **176**, 131–167 (2009)
25. Hochman, M., Meyerovitch, T.: A characterization of the entropies of multidimensional shifts of finite type. Ann. Math. **171**, 2011–2038 (2010)
26. Jeandel, E., Vanier, P.: A characterization of subshifts with computable language. In: STACS Art. No. 40, LIPIcs. Leibniz International Proceedings in Informatics, 126, Schloss Dagstuhl. Leibniz-Zentrum für Informatik, Wadern (2019)

27. Kannan, R., Lipton, R.J.: Polynomial-time algorithm for the orbit problem. J. Assoc. Comput. Mach. **33**(4), 808–821 (1986)
28. Keane, M.: Non-ergodic interval exchange transformations. Israel J. Math. **26**, 188–196 (1977)
29. Lagarias, J.C.: The quality of the Diophantine approximations found by the Jacobi-Perron algorithm and related algorithms. Monatsh. Math. **115**, 299–328 (1993)
30. Lhote, L.: Computation of a class of continued fraction constants. In: Analytic Algorithmic and Combinatorics ANALCO 2014, pp. 199–210. SIAM (2004)
31. Lothaire, M.: Algebraic Combinatorics on Words. Encyclopedia of Mathematics and Its Applications, vol. 90. Cambridge University Press, Cambridge (2002)
32. Morse, M., Hedlund, G.A.: Symbolic dynamics II. Sturmian Trajectories Am. J. Math. **62**, 1–42 (1940)
33. Pollicott, M.: Maximal Lyapunov exponents for random matrix products. Invent. Math. **181**, 209–226 (2010)
34. Priebe Frank, N., Sadun, L.: Fusion: a general framework for hierarchical tilings of $\mathbb{R}^d$. Geometriae Dedicata **171**, 149–186 (2014)
35. Pytheas, N.: Substitutions in dynamics, arithmetics and combinatorics. In: Berthé, V., Ferenczi, S., Mauduit, C., Siegel, A. (eds.) Lecture Notes in Mathematics, vol. 1794. Springer, Heidelberg (2002)
36. Queffélec, M.: Substitution Dynamical Systems—Spectral Analysis. Lecture Notes in Mathematics, vol. 1294, 2nd edn. Springer, Berlin (2010). https://doi.org/10.1007/978-3-642-11212-6
37. Robinson, R.M.: Undecidability and nonperiodicity for tilings of the plane. Invent. Math. **12**, 177–209 (1971)
38. Schweiger, F.: Multidimensional Continued Fractions. Oxford Science Publications, Oxford University Press, Oxford (2000)
39. Seneta, E.: Non-Negative Matrices and Markov Chains. Springer Series in Statistics. Springer, New York (1981)
40. Siegel, A. Thuswaldner, J.M.: Topological properties of Rauzy fractals. Mém. Soc. Math. Fr. **118** (2009)
41. Shallit, J.: Decidability and enumeration for automatic sequences: a survey. In: Bulatov, A.A., Shur, A.M. (eds.) CSR 2013. LNCS, vol. 7913, pp. 49–63. Springer, Heidelberg (2013). https://doi.org/10.1007/978-3-642-38536-0_5
42. Tsitsiklis, J.N., Blondel, J.N.: The Lyapunov exponent and joint spectral radius of pairs of matrices are hard—when not impossible—to compute and to approximate. Math. Control. Signals Syst. **10**, 31–40 (1997)
43. Yoccoz, J.-C.: Continued fraction algorithms for interval exchange maps: an introduction. In: Frontiers in Number Theory, Physics, and Geometry I, pp. 401–435. Springer, Berlin (2006)
44. Zorich, A.: Deviation for interval exchange transformations. Ergod. Theory Dyn. Syst. **17**, 1477–1499 (1997)

Games with Full, Longitudinal, and Transverse Observability

Orna Kupferman[✉]

School of Computer Science and Engineering, The Hebrew University,
Jerusalem, Israel
orna@cs.huji.ac.il

Abstract. Design and control of multi-agent systems correspond to the synthesis of winning strategies in games that model the interaction between the agents. In games with *full observability*, the strategies of players depend on the full history of the play. In games with partial observability, strategies depend only on observable components of the history. We survey two approaches to partial observability in two-player turn-based games with behavioral winning conditions. The first is the traditional *longitudinal observability*, where in all vertices, the players observe the assignment only to an observable subset of the atomic propositions. The second is the recently studied *transverse observability*, where players observe the assignment to all the atomic propositions, but only in vertices they own.

1 Introduction

Design and control of multi-agent systems correspond to the synthesis of winning strategies in games that model the interaction between the agents. The game is played on a graph whose paths correspond to computations of the system. We study here settings in which each of the players has control in different parts of the system. Thus, the game is *turn-based*: The set V of vertices is partitioned between the players. In the beginning of a *play* in the game, a token is placed on an initial vertex. Then, in each turn, the player that owns the vertex that hosts the token chooses a successor vertex and moves there the token. The objectives of the players refer to the infinite play that they generate. We consider *behavioral* objectives: the vertices in V are labeled by assignments to a set AP of atomic propositions – these with respect to which the system is defined, and the winning condition is a language L of infinite words in $(2^{AP})^\omega$. One player aims for a play whose computation is in L, while her opponent aims for a play whose computation is not in L. A (deterministic) *strategy* for a player directs her how to continue a play that reaches her vertices.

Early work considers games with *full observability*, where strategies depend on the full history of the play. In contrast, in games with *partial observability*, strategies depend only on visible components of the history. We survey

O.K—Supported in part by the Israel Science Foundation, grant No. 2357/19.

S. Schmitz and I. Potapov (Eds.): RP 2020, LNCS 12448, pp. 20–34, 2020.
https://doi.org/10.1007/978-3-030-61739-4_2

two approaches to partial observability. The first is the traditional *longitudinal observability*, where in all vertices, the players observe the assignment only to an observable subset of the atopic propositions [3–5,10,17]. The second is the recently studied *transverse observability*, where players observe the assignment to all the atomic propositions, but only in the vertices they own [9].

Both types of partial observability correspond to realistic settings. In games with longitudinal observability, we capture systems in which each of the underlying components can only view and control a subset of the system's variables. For example, a program that interacts with an environment with private variables. In games with transverse observability, we capture systems in which control is switched among the underlying components, which can observe only these parts of the interaction that they control. For example, a communication network in which a company that owns part of the routers has to make routing decisions based only on information about visits to its routers [1], a composite reactive system that does not observe the interaction of the environment with the other components [13], and switched systems where components are activated by a scheduler and are not aware of the evolution of the system while being switched off [6,12,14,16].

Technically, in games with longitudinal observability, we assume that each player can observe a subset $O \subseteq AP$ of the atomic propositions, and her strategy depends only on the history of the assignments to O. Accordingly, players cannot distinguish between different histories of the game that differ in their assignments to the hidden atomic propositions. Note that the uncertainty of a player when she moves the token concerns the vertex in V in which the token is – each observable history actually corresponds to a set of vertices – all these that are reachable via the observable history. Moreover, rather than directing the token to a successor vertex, the strategy can only suggest an assignment to the observable atomic propositions.

Then, in games with transverse observability, the strategy for a player cannot distinguish among histories that differ in visits to vertices owned by other players. Consequently, the uncertainty of a player when she moves the token concerns the computation traversed by the token so far – each observable history may correspond to different computations – all these that agree on the *perspective* view of the player, namely these whose restriction to vertices owned by the player coincide.[1]

We show that *universal tree automata* are a suitable tool for handling both types of partial observability. We reduce the problem of deciding whether a player wins a game to the non-emptiness problem of a universal tree automaton that accepts all winning strategies for the player. Using universal automata, we can

---

[1] Note that *memoryless* strategies, which depend only on the current vertex of the game, are a special case of strategies with a transverse observability. Since, however, we consider here games with a *behavioral* winning condition, memoryless strategies are not sufficient. In contrast, in some games with a *structural* winning condition, say Büchi or parity, memoryless strategies are sufficient, and hence also ones with transverse observability.

guarantee that the strategy behave in the same way in indistinguishable positions. Technically, in the longitudinal model, where uncertainty concern vertices in V, this amounts to the universal automaton sending copies to nodes in the tree that correspond to different vertices in V. Accordingly, the complexity is exponential in the size of the game.[2] Then, in the transverse model, this amounts to the universal automaton sending copies to nodes in the tree that correspond to different states in the automaton for the winning condition. Accordingly, the complexity is only polynomial in the game and exponential in the winning condition. In Sect. 4, we discuss the complete picture of the complexity in the different settings.

2   Preliminaries

2.1   Games

A *game graph* is a tuple $G = \langle AP, V_1, V_2, v_0, E, \tau \rangle$, where AP is a finite set of atomic propositions, V_1 and V_2 are disjoint sets of vertices, owned by PLAYER 1 and PLAYER 2, respectively, and we let $V = V_1 \cup V_2$. Then, $v_0 \in V_1$ is an initial vertex, which we assume to be owned by PLAYER 1, and $E \subseteq V \times V$ is a total edge relation, thus for every $v \in V$ there is $v' \in V$ such that $\langle v, v' \rangle \in E$. The function $\tau : V \to 2^{AP}$ maps each vertex to a set of atomic propositions that hold in it. We assume that vertices are uniquely labeled. Thus, if $v \neq v'$, then $\tau(v) \neq \tau(v')$. Note that if this is not the case, we can add to AP atomic propositions that encode V. The *size* $|G|$ of G is $|E|$, namely the number of edges in it.

In the beginning of a play in the game, a token is placed on v_0, Then, in each turn, the player that owns the vertex that hosts the token chooses a successor vertex and moves there the token. A *play* $\rho = v_0, v_1, \ldots$ in G, is an infinite path in G that starts in v_0; thus $\langle v_i, v_{i+1} \rangle \in E$ for all $i \geq 0$. The play ρ induces a *computation* $\tau(\rho) = \tau(v_0), \tau(v_1), \ldots \in (2^{AP})^\omega$. A *game* is a pair $\mathcal{G} = \langle G, L \rangle$, where G is a game graph, and $L \subseteq (2^{AP})^\omega$ is a *behavioral winning condition*, namely an ω-regular language over the atomic propositions, given by an LTL formula or an automaton. Intuitively, PLAYER 1 aims for a play whose computation is in L, while PLAYER 2 aims for a play whose computation is in $comp(L) = (2^{AP})^\omega \setminus L$.

Let $\mathsf{Prefs}(G)$ be the set of nonempty prefixes of plays in G. For a finite sequence $\rho = v_0, \ldots, v_n$ of vertices, let $\mathsf{Last}(\rho) = v_n$. For $j \in \{1, 2\}$, let $\mathsf{Prefs}_j(G) = \{\rho \in \mathsf{Prefs}(G) : \mathsf{Last}(\rho) \in V_j\}$. Thus, $\mathsf{Prefs}_j(G)$ is the subset of $\mathsf{Prefs}(G)$ consisting of prefixes of plays whose last vertex is in V_j. A *strategy* for PLAYER j is a function $f_j : \mathsf{Prefs}_j(G) \to V$ such that for every $\rho \in \mathsf{Prefs}_j(G)$, we have that $\langle \mathsf{Last}(\rho), f_j(\rho) \rangle \in E$. That is, a strategy for PLAYER j maps prefixes of plays that end in a vertex v she owns to a successor of v. For technical convenience, we add ϵ to $\mathsf{Prefs}_1(G)$ and require all strategies f_1 of PLAYER 1 to have $f_1(\epsilon) = v_0$. Thus, all plays start with PLAYER 1 placing the token on v_0.

---

[2] It is important to note that the complexities studies here consider *two-player* games. For settings with more players, the problem is undecidable [3,15,17].

The *outcome* of two strategies f_1 and f_2 of PLAYER 1 and PLAYER 2, respectively, is the play obtained when the players follow the strategies f_1 and f_2. Formally, $\mathsf{Outcome}(f_1, f_2) = v_0, v_1, \ldots$ is such that for all $i \geq 0$, if $v_i \in V_j$, then $v_{i+1} = f_j(v_0, \ldots, v_i)$.

The above definition assumes that both players have *full observability* of the play generated by their strategies. Indeed, the domain of a strategy is the full history of the play so far. We consider two types of partial observability: *longitudinal*, where in all vertices the players observe the assignment only to an observable subset of the atopic propositions, and *transverse*, where players observe the assignment to all the atomic propositions, but only in the vertices they own.

We start with games with longitudinal observability. We assume that the set AP of atomic propositions is partitioned into two sets O and H, of *observable* and *hidden* atomic propositions. Thus, $AP = O \cup H$, with $O \cap H = \emptyset$. We note that in many settings, the partition to observable and hidden atomic propositions is parameterized by the players; thus PLAYER 1 and PLAYER 2 observe different sets of atomic propositions. It is easy to extend our results here to these settings. Let $\tau_O : V \to 2^O$ and $\tau_H : V \to 2^H$ be the projections of τ on O and H, respectively; thus, for every $v \in V$, we have that $\tau_O(v) = \tau(v) \cap O$ and $\tau_H(v) = \tau(v) \cap H$. The transitions of G and the partition of V to V_1 and V_2 depend only on the atomic propositions in O. Formally, there is a function $E_O \subseteq 2^O \times 2^O$ such that for all $v, v' \in V$, we have that $\langle v, v' \rangle \in E$ iff $\langle \tau_O(v), \tau_O(v') \rangle \in E_O$. Also, there are sets $\Theta_1, \Theta_2 \subseteq 2^O$ such that $V_1 = \tau_O^{-1}(\Theta_1)$ and $V_2 = \tau_O^{-1}(\Theta_2)$.

In games with longitudinal observability, the players see only the assignments to the observable atomic propositions. Every play $\rho = v_0, v_1, \ldots$ in G induces an *observable computation* $\tau_O(\rho) = \tau_O(v_0), \tau_O(v_1), \ldots \in (2^O)^\omega$. For a sequence $o = o_0, \ldots, o_n$ of assignments to atomic propositions in O, thus $o \in (2^O)^*$, let $\mathsf{Last}(o) = o_n$. Let $\mathsf{OPrefs}(G)$ be the set of nonempty prefixes of observable computations in G. For $j \in \{1, 2\}$, let $\mathsf{OPrefs}_j(G) = \{o \in \mathsf{OPrefs}(G) : \mathsf{Last}(o) \in \Theta_j\}$. Thus, $\mathsf{OPrefs}_j(G)$ is the subset of $\mathsf{OPrefs}(G)$ consisting of prefixes of observable computations whose last assignment is in Θ_j. Note that, equivalently, $\mathsf{OPrefs}_j(G)$ is the subset of $\mathsf{OPrefs}(G)$ consisting of prefixes of observable computations induced by prefixes $\rho \in \mathsf{Prefs}_j(G)$ with $\mathsf{Last}(\rho) \in V_j$.

Recall that the players see only the assignments to the atomic propositions in O. Accordingly, a strategy for PLAYER j is a function $f_j : \mathsf{OPrefs}_j(G) \to 2^O$ such that for every $o \in \mathsf{OPrefs}_j(G)$, we have that $\langle \mathsf{Last}(o), f_j(o) \rangle \in E_O$. That is, a strategy for PLAYER j maps prefixes of observable computations that end in an assignment $\theta \in \Theta_j$ to an assignment θ' such that $E_O(\theta, \theta')$. As in the case of full observability, we add ϵ to $\mathsf{OPrefs}_1(G)$ and require all strategies f_1 of PLAYER 1 to have $f_1(\epsilon) = \tau_O(v_0)$. The outcome of two strategies f_1 and f_2 of PLAYER 1 and PLAYER 2, respectively, is then a set of plays – all these that may be traversed when the players follow the strategies f_1 and f_2. Formally, $v_0, v_1, v_2 \ldots \in \mathsf{Outcome}(f_1, f_2)$ is such that for all $i \geq 0$, if $v_i \in V_j$, then $\tau_O(v_{i+1}) = f_j(\tau_O(v_0), \ldots, \tau_O(v_i))$. Note that the different plays in $\mathsf{Outcome}(f_1, f_2)$ agree on the observable computation they induce and differ only by the assignments to atomic propositions in H. Note also that full observability can be viewed as a special case of longitudinal observability, with $H = \emptyset$.

We continue to games with transverse observability, where the observability of PLAYER j is restricted to vertices she owns. For a prefix $\rho = v_0, \ldots, v_i \in$ Prefs(G) and $j \in \{1, 2\}$, *the perspective of* PLAYER j *on* ρ, denoted Persp$_j(\rho)$, is the restriction of ρ to vertices $v_i \in V_j$. We denote the perspectives of PLAYER j on prefixes in Prefs$_j(G)$ by PPrefs$_j(G)$. Formally, PPrefs$_j(G) = \{$Persp$_j(\rho) : \rho \in$ Prefs$_j(G)\}$. Note that PPrefs$_j(G) \subseteq V_j^*$. A *strategy* for PLAYER j is then a function $f_j :$ PPrefs$_j(G) \rightarrow V$ such that for every $\rho \in$ PPrefs$_j(G)$, we have that $\langle$Last$(\rho), f_j(\rho)\rangle \in E$. That is, a strategy for PLAYER j maps her perspective of prefixes of plays that end in a vertex v she owns to a successor of v. The *outcome* of two strategies f_1 and f_2 of PLAYER 1 and PLAYER 2, respectively, is the play obtained when the players follow the strategies f_1 and f_2. Formally, Outcome$(f_1, f_2) = v_0, v_1, \ldots$ is such that for all $i \geq 0$, if $v_i \in V_j$, then $v_{i+1} = f_j($Persp$_j(v_0, \ldots, v_i))$.

Consider a game $\mathcal{G}$. For games with full or transverse observability, we say that PLAYER 1 *wins* $\mathcal{G}$ if there is a strategy f_1 for PLAYER 1 such that for every strategy f_2 for PLAYER 2, we have that $\tau($Outcome$(f_1, f_2)) \in L$. For games with longitudinal observability, the outcome of two strategies may be several plays. Then, PLAYER 1 *wins* $\mathcal{G}$ if there is a strategy f_1 for PLAYER 1 such that for every strategy f_2 for PLAYER 2 and play $\rho \in$ Outcome$(f_1, f_2))$, we have that $\tau(\rho) \in L$.

2.2   Automata

Solving games with incomplete information, we are going to use automata on infinite trees, defined below.

Given a set D of directions, a *D-tree* is a set $T \subseteq D^*$ such that if $x \cdot c \in T$, where $x \in D^*$ and $c \in D$, then also $x \in T$. The elements of T are called *nodes*, and the empty word ε is the *root* of T. For every $x \in T$, the nodes $x \cdot c$, for $c \in D$, are the *successors* of x. A *path* π of a tree T is a set $\pi \subseteq T$ such that $\varepsilon \in \pi$ and for every $x \in \pi$, either x is a leaf or there exists a unique $c \in D$ such that $x \cdot c \in \pi$. Given an alphabet Σ, a *Σ-labeled D-tree* is a pair $\langle T, \tau \rangle$ where T is a tree and $\tau : T \rightarrow \Sigma$ maps each node of T to a letter in Σ.

For a set X, let $\mathcal{B}^+(X)$ be the set of positive Boolean formulas over X (i.e., Boolean formulas built from elements in X using $\land$ and $\lor$), where we also allow the formulas *True* and *False*. For a set $Y \subseteq X$ and a formula $\theta \in \mathcal{B}^+(X)$, we say that Y *satisfies* θ iff assigning *True* to elements in Y and assigning *False* to elements in $X \setminus Y$ makes θ true. An *alternating tree automaton* is $\mathcal{A} = \langle \Sigma, D, Q, q_{in}, \delta, \alpha \rangle$, where Σ is the input alphabet, D is a set of directions, Q is a finite set of states, $\delta : Q \times \Sigma \rightarrow \mathcal{B}^+(D \times Q)$ is a transition function, $q_{in} \in Q$ is an initial state, and α specifies the acceptance condition (a condition that defines a subset of Q^ω; we define several types of acceptance conditions below). For a state $q \in Q$, we use $\mathcal{A}^q$ to denote the automaton obtained from $\mathcal{A}$ by setting the initial state to be q. The *size* of $\mathcal{A}$, denoted $|\mathcal{A}|$, is the sum of lengths of formulas that appear in δ.

The alternating automaton $\mathcal{A}$ runs on Σ-labeled D-trees. A *run* of $\mathcal{A}$ over a Σ-labeled D-tree $\langle T, \tau \rangle$ is a $(T \times Q)$-labeled $\mathbb{N}$-tree $\langle T_r, r \rangle$. Each node of T_r corresponds to a node of T. A node in T_r, labeled by (x, q), describes a copy of the automaton that reads the node x of T and visits the state q. Note that many nodes of T_r can correspond to the same node of T. The labels of a node and its successors have to satisfy the transition function. Formally, $\langle T_r, r \rangle$ satisfies the following:

1. $\varepsilon \in T_r$ and $r(\varepsilon) = \langle \varepsilon, q_{in} \rangle$.
2. Let $y \in T_r$ with $r(y) = \langle x, q \rangle$ and $\delta(q, \tau(x)) = \theta$. Then there is a (possibly empty) set $S = \{(c_0, q_0), (c_1, q_1), \ldots, (c_{n-1}, q_{n-1})\} \subseteq D \times Q$, such that S satisfies θ, and for all $0 \le i \le n-1$, we have $y \cdot i \in T_r$ and $r(y \cdot i) = \langle x \cdot c_i, q_i \rangle$.

For example, if $\langle T, \tau \rangle$ is a $\{0, 1\}$-tree with $\tau(\varepsilon) = a$ and $\delta(q_{in}, a) = ((0, q_1) \vee (0, q_2)) \wedge ((0, q_3) \vee (1, q_2))$, then, at level 1, the run $\langle T_r, r \rangle$ includes a node labeled $(0, q_1)$ or a node labeled $(0, q_2)$, and includes a node labeled $(0, q_3)$ or a node labeled $(1, q_2)$. Note that if, for some y, the transition function δ has the value *True*, then y need not have successors. Also, δ can never have the value *False* in a run.

A run $\langle T_r, r \rangle$ is accepting if all its infinite paths satisfy the acceptance condition. Given a run $\langle T_r, r \rangle$ and an infinite path $\pi \subseteq T_r$, let $inf(\pi) \subseteq Q$ be such that $q \in inf(\pi)$ if and only if there are infinitely many $y \in \pi$ for which $r(y) \in T \times \{q\}$. That is, $inf(\pi)$ contains exactly all the states that appear infinitely often in π. We consider here three acceptance conditions defined as follows. A path π satisfies:

- a *Büchi* condition $\alpha \subseteq Q$ iff $inf(\pi) \cap \alpha \neq \emptyset$.
- a *co-Büchi* condition $\alpha \subseteq Q$ iff $inf(\pi) \cap \alpha = \emptyset$.
- a *parity* condition $\alpha : Q \to \{1, \ldots, k\}$ iff the minimal color $i \in \{1, \ldots, k\}$ for which $inf(\pi) \cap \alpha^{-1}(i) \neq \emptyset$, is even. The number k of colors in α is called the *index* of the automaton.

For the three conditions, an automaton accepts a tree iff there exists a run that accepts it. We denote by $L(\mathcal{A})$ the set of all Σ-labeled D-trees that $\mathcal{A}$ accepts.

Below we discuss some special cases of alternation automata. The alternating automaton $\mathcal{A}$ is *nondeterministic* if for all the formulas that appear in δ, if (c_1, q_1) and (c_2, q_2) are conjunctively related, then $c_1 \neq c_2$. (i.e., if the transition is rewritten in disjunctive normal form, there is at most one element of $\{c\} \times Q$, for each $c \in D$, in each disjunct). The automaton $\mathcal{A}$ is *universal* if all the formulas that appear in δ are conjunctions of atoms in $D \times Q$, and $\mathcal{A}$ is *deterministic* if it is both nondeterministic and universal. The automaton $\mathcal{A}$ is a *word* automaton if $|D| = 1$. Then, we can omit D from the specification of the automaton and denote the transition function of $\mathcal{A}$ as $\delta : Q \times \Sigma \to \mathcal{B}^+(Q)$. If the word automaton is nondeterministic or universal, then $\delta : Q \times \Sigma \to 2^Q$, and we often extend δ to sets of states and to finite words: for $S \subseteq Q$, we have that $\delta(S, \epsilon) = S$ and for a word $w \in \Sigma^*$ and a letter $\sigma \in \Sigma$, we have $\delta(S, w \cdot \sigma) = \delta(\delta(S, w), \sigma)$.

We sometimes refer also to automata on finite words. There, $\alpha \subseteq Q$ and a (finite) run is accepting if its last state is in α. Finally, we say that a Σ-labeled D-tree $\langle T, \tau \rangle$ is *regular* if for all letters $\sigma \in \Sigma$, we have that $\tau^{-1}(\sigma)$ is a regular language over D. Note that a regular tree can be generated by a (D, Σ)-*transducer*: a deterministic automaton over D in which each state is labeled by a letter in Σ. Then, $\tau(x)$, for a node $x \in D^*$, is the letter that labels the transducer state that is reachable by reading x.

We denote each of the different types of automata by three-letter acronyms in $\{D, N, U, A\} \times \{F, B, C, P\} \times \{W, T\}$, where the first letter describes the branching mode of the automaton (deterministic, nondeterministic, universal, or alternating), the second letter describes the acceptance condition (finite, Büchi, co-Büchi, or parity), and the third letter describes the object over which the automaton runs (words or trees). For example, APT are alternating parity tree automata and UCT are universal co-Büchi tree automata.

3 Deciding Games

In this section we describe upper bounds to the problem of deciding whether PLAYER 1 wins a given game. In the three types of observability, we reduce the problem to the nonemptiness problem of a UCT that accepts strategies that are winning for PLAYER 1. The solutions are similar, and the technical differences highlight the conceptual difference among the two types of partial observability. On the one hand, longitudinal observability induces uncertainty about the vertex that hosts of the token, and requires algorithms that are exponential in the size of the game graph. On the other hand, transverse observability induces uncertainty about the computation traversed so far, and requires algorithms that are exponential in the size of an automaton that specifies the winning condition.

3.1 Games with Full Observability

In a game with full observability, we model strategies by $(V \cup \{\oslash\})$-labeled V-trees. In nodes in V^* that correspond to prefixes of plays in $\mathsf{Prefs}_1(G)$, a strategy for PLAYER 1 should return a vertex to which the token moves. In other nodes, it should return $\oslash$, to indicate that PLAYER 1 does not move. In more details, a strategy for PLAYER 1 is $f_1 : V^* \to V \cup \{\oslash\}$ such that $f_1(\epsilon) = v_0$, for all $\rho \in \mathsf{Prefs}_1(G) \setminus \{\epsilon\}$, we have that $\langle \mathsf{Last}(\rho), f_1(\rho) \rangle \in E$, and for all $\rho \in V^* \setminus \mathsf{Prefs}_1(G)$, we have $f_1(\rho) = \oslash$.

Consider a game $\mathcal{G} = \langle G, L \rangle$. Let $G = \langle AP, V_1, V_2, E, v_0, \tau \rangle$, and assume that L is given by a UCW $\mathcal{U} = \langle 2^{AP}, Q, q_0, \delta, \alpha \rangle$. Suppose that the token is placed in some vertex v and that the objective of PLAYER 1 is to force the token into computations in $L(\mathcal{U}^q)$, for a state $q \in Q$. In particular, in the beginning of every play, the token is placed on v_0 and the objective of PLAYER 1 is to force the token into computations in $L(\mathcal{U}^{q_0})$. If $v \in V_1$, then PLAYER 1 chooses to move the token to some successor v' of v. Then, the new objective of PLAYER 1 is to force the token from v' into computations in $L(\mathcal{U}^{q'})$, for all the states

$q' \in \delta(q, \tau(v))$. If $v \in V_2$, then PLAYER 2 may choose to move the token to any successor v' of v. Then, the new objective of PLAYER 1 is to force the token from all successors v' of v into computations in $L(\mathcal{U}^{q'})$, for all the states $q' \in \delta(q, \tau(v))$.

The above intuition motivates the following definition of *updated objectives*. For a pair $\langle v, q \rangle \in V \times Q$, standing for an objective of PLAYER 1 to force a token placed on v to be accepted from q, and an action $\sigma \in V \cup \{\oslash\}$ of PLAYER 1, we define the set $S^\sigma_{v,q} \subseteq (V \times Q) \cup \{False\}$ of updated objectives – these that PLAYER 1 has to satisfy in order to fulfil her $\langle v, q \rangle$ objective after choosing σ. Formally, we define $S^\sigma_{v,q}$ as follows.

- If $v \in V_2$ and $\sigma = \oslash$, then $S^\sigma_{v,q} = \{\langle v', q' \rangle : E(v, v')$ and $q' \in \delta(q, \tau(v))\}$.
- If $v \in V_1$ and $\sigma = v'$ for $v' \in V$ with $E(v, v')$, then $S^\sigma_{v,q} = \{\langle v', q' \rangle : q' \in \delta(q, \tau(v))\}$.
- Otherwise, $S^\sigma_{v,q} = \{False\}$.

We can now use the notion of updated objectives in order to define a UCT for winning strategies:

Theorem 1. *Let $\mathcal{G} = \langle G, \mathcal{U} \rangle$ be a full-observability game, where G is a game graph and $\mathcal{U}$ is a UCW. We can construct a UCT $\mathcal{A}_\mathcal{G}$ over $(V \cup \{\oslash\})$-labeled V-trees such that $\mathcal{A}_\mathcal{G}$ accepts a $(V \cup \{\oslash\})$-labeled V-tree $\langle V^*, \eta \rangle$ iff $\langle V^*, \eta \rangle$ is a winning strategy for PLAYER 1. The size of $\mathcal{A}_\mathcal{G}$ is polynomial in $|G|$ and $|\mathcal{U}|$.*

Proof. Let $\mathcal{U} = \langle 2^{AP}, Q, q_0, \delta, \alpha \rangle$. We define $\mathcal{A}_\mathcal{G} = \langle V \cup \{\oslash\}, V, Q', q'_0, \delta', \alpha' \rangle$, where

- $Q' = V \times Q$. Intuitively, when $\mathcal{A}$ is in state $\langle v, q \rangle$, it accepts strategies that force a token placed on v into a computation accepted by $\mathcal{U}^q$.
- $q'_0 = \langle v_0, q_0 \rangle$
- For all $\langle v, q \rangle \in V \times Q$ and $\sigma \in V \cup \{\oslash\}$, if $S^\sigma_{v,q} = \{False\}$, then $\delta'(\langle v, q \rangle, \sigma) = False$. Otherwise,

$$\delta'(\langle v, q \rangle, \sigma) = \bigwedge_{\langle v', q' \rangle \in S^\sigma_{v,q}} (v', \langle v', q' \rangle).$$

Thus, for every updated objective $\langle v', q' \rangle \in S^\sigma_{v,q}$, the automaton sends a copy in state $\langle v', q' \rangle$ to direction v'. Note that since $\mathcal{U}$ is universal, it may send copies in different states to the same direction. Note, however, that all copies sent to direction v' agree on their V-component, which is v'. Note also that when $S^\sigma_{v,q} = \emptyset$, we get that $\delta'(\langle v, q \rangle, \sigma) = True$. Finally, note that if $v \in V_1$, then all copies are sent to one directions, whereas if $v \in V_2$, they are sent to all directs v' with $E(v, v')$.
- $\alpha' = V \times \alpha$.

Intuitively, the branches of the run tree of $\mathcal{A}_\mathcal{G}$ over $\langle V^*, \eta \rangle$ correspond to all the possible runs of $\mathcal{U}$ on all the plays into which PLAYER 1 can force the game when she follows the strategy η. All these runs have to be accepted, which is guaranteed by the definition of $\mathcal{A}_\mathcal{G}$. $\square$

Theorem 2. *Let* $\mathcal{G} = \langle G, \mathcal{U} \rangle$ *be a full-observability game, where* G *is a game graph and* $\mathcal{U}$ *is a UCW. We can construct an NBT* $\mathcal{A}'_{\mathcal{G}}$ *over* $(V \cup \{\oslash\})$*-labeled* V*-trees such that there is a winning strategy for* PLAYER 1 *in* $\mathcal{G}$ *iff* $L(\mathcal{A}'_{\mathcal{G}})$ *is not empty. The size of* $\mathcal{A}'_{\mathcal{G}}$ *is polynomial in* $|G|$ *and exponential in* $|\mathcal{U}|$.

Proof. By Theorem 1, we can construct a UCT $\mathcal{A}_{\mathcal{G}}$ over $(V \cup \{\oslash\})$-labeled V-trees such that $L(\mathcal{A}_{\mathcal{G}})$ is not empty iff there is a winning strategy for PLAYER 1 in $\mathcal{G}$. The size of $\mathcal{A}_{\mathcal{G}}$ is polynomial in $|G|$ and $|\mathcal{U}|$. The transformation from $\mathcal{A}_{\mathcal{G}}$ to $\mathcal{A}'_{\mathcal{G}}$ can be done by the method of [11]. Below we analyze the construction and show how the fact that $\mathcal{A}_{\mathcal{G}}$ is deterministic in the V-component implies that it is exponential in $|\mathcal{U}|$ and only polynomial in $|G|$.

For $k \geq 1$, let $[k] = \{1, \ldots, k\}$. The construction in [11] transforms the UCT $\mathcal{A}_{\mathcal{G}} = \langle V \cup \{\oslash\}, V, Q', q'_0, \delta', \alpha' \rangle$ to an NBT $\mathcal{A}'_{\mathcal{G}}$ with states $S = 2^{Q' \times [k]} \times 2^{Q' \times [k]}$, where k is such that $|Q'| \cdot k$ bounds the size of an NRT that is equivalent to $\mathcal{A}_{\mathcal{G}}$, which is exponential in $|Q'|$. Also, for every state $\langle P, O \rangle \in S$, we have $O \subseteq P$, and if $\langle p, i \rangle$ and $\langle p', i' \rangle$ are in P with $p = p'$, then $i = i'$. Therefore, the states in S can be written as $2^S \times 2^S \times \mathcal{F}$, where $\mathcal{F}$ is the set of functions $f : Q' \to [k]$. Recall that the states of the UCT $\mathcal{A}_{\mathcal{G}}$ are $Q' = V \times Q$, and that $\mathcal{A}_{\mathcal{G}}$ is deterministic in the V-component. Hence, the translation of $\mathcal{A}_{\mathcal{G}}$ to an NRT is polynomial in $|G|$ and exponential in $|\mathcal{U}|$, and thus k is only polynomial in $|G|$. Also, for every $\langle P, O \rangle \in S$, if $\langle v, q, i \rangle$ and $\langle v', q', i' \rangle$ are in P, then since $\mathcal{A}_{\mathcal{G}}$ is deterministic in the V-component, we have $v = v'$. Therefore, the states in S can be written as $V \times 2^Q \times 2^Q \times \mathcal{F}$, where $\mathcal{F}$ is the set of functions $f : Q \to [k]$. Hence, $|S|$ is polynomial in $|G|$ and exponential in $|\mathcal{U}|$. $\qquad\square$

Since the nonemptiness problem for an NBT $\mathcal{A}$ can be solved in quadratic time [19], and we can return a transducer of size $O(|\mathcal{A}|)$ that witnesses the nonemptiness, we can conclude with the following upper bound:

Corollary 1. *Deciding whether* PLAYER 1 *wins in a game* $\langle G, \mathcal{U} \rangle$ *for a UCW* $\mathcal{U}$ *is in EXPTIME. The problem can be solved in time polynomial in* $|G|$ *and exponential in* $|\mathcal{U}|$. *Moreover, when* PLAYER 1 *wins, the algorithm returns a witness strategy by means of a transducer of size polynomial in* $|G|$ *and exponential in* $|\mathcal{U}|$.

3.2   Games with Longitudinal Observability

In a game with longitudinal observability, we model strategies by $(2^O \cup \{\oslash\})$-labeled 2^O-trees. In nodes in $(2^O)^*$ that correspond to prefixes of observable computations in $\mathsf{OPrefs}_1(G)$, a strategy for PLAYER 1 should return an assignment to the observable atomic propositions, to which the token moves. In other nodes, it should return $\oslash$, to indicate that PLAYER 1 does not move. In more details, a strategy for PLAYER 1 is $f_1 : (2^O)^* \to 2^O \cup \{\oslash\}$ such that $f_1(\epsilon) = \tau_O(v_0)$, for all $o \in \mathsf{OPrefs}_1(G) \setminus \{\epsilon\}$, we have that $\langle \mathsf{Last}(o), f_1(o) \rangle \in E_O$, and for all $o \in (2^O)^* \setminus \mathsf{OPrefs}_1(G)$, we have $f_1(o) = \oslash$.

Consider a game $\mathcal{G} = \langle G, L \rangle$. Let $G = \langle AP, V_1, V_2, E, v_0, \tau \rangle$, and assume that L is given by a UCW $\mathcal{U} = \langle 2^{AP}, Q, q_0, \delta, \alpha \rangle$. Suppose that the token is placed in some vertex v and that the objective of PLAYER 1 is to force the token into computations in $L(\mathcal{U}^q)$, for a state $q \in Q$. Note that PLAYER 1 does not know v, she only knows the prefix of the observable computation that leads to v. Accordingly, the definition of *updated objectives* is now as follows. For a pair $\langle v, q \rangle \in V \times Q$, standing for an objective of PLAYER 1 to force a token placed on v to be accepted from q, and an action $\sigma \in 2^O \cup \{\oslash\}$ of PLAYER 1, we define the set $S_{v,q}^\sigma \subseteq (V \times Q) \cup \{False\}$ of updated objectives – these that PLAYER 1 has to satisfy in order to fulfil her $\langle v, q \rangle$ objective after choosing σ. Formally, we define $S_{v,q}^\sigma$ as follows.

- If $v \in V_2$ and $\sigma = \oslash$, then $S_{v,q}^\sigma = \{\langle v', q' \rangle : E(v, v') \text{ and } q' \in \delta(q, \tau(v))\}$.
- If $v \in V_1$ and $\sigma = \theta'$ for $\theta' \in 2^O$ with $E_O(\theta, \theta')$, then $S_{v,q}^\sigma = \{\langle v', q' \rangle : E(v, v'), \tau_O(v') = \theta', \text{ and } q' \in \delta(q, \tau(v))\}$.
- Otherwise, $S_{v,q}^\sigma = \{False\}$.

The construction of a UCT for winning strategies is then similar to the one described in Theorem 1, adjusted to directions in 2^O and the above definition of updated objectives.

Theorem 3. *Let $\mathcal{G} = \langle G, \mathcal{U} \rangle$ be a game with longitudinal observability, where G is a game graph and $\mathcal{U}$ is a UCW. We can construct a UCT $\mathcal{A}_\mathcal{G}$ over $(2^O \cup \{\oslash\})$-labeled 2^O-trees such that $\mathcal{A}_\mathcal{G}$ accepts a $(2^O \cup \{\oslash\})$-labeled V-tree $\langle V^*, \eta \rangle$ iff $\langle V^*, \eta \rangle$ is a winning strategy for PLAYER 1. The size of $\mathcal{A}_\mathcal{G}$ is polynomial in $|G|$ and $|\mathcal{U}|$.*

Proof. Let $\mathcal{U} = \langle 2^{AP}, Q, q_0, \delta, \alpha \rangle$. We define the UCT $\mathcal{A}_\mathcal{G} = \langle V \cup \{\oslash\}, V, V \times Q, \langle v_0, q_0 \rangle, \delta', V \times \alpha \rangle$, where for all $\langle v, q \rangle \in V \times Q$ and $\sigma \in 2^O \cup \{\oslash\}$, if $S_{v,q}^\sigma = \{False\}$, then $\delta'(\langle v, q \rangle, \sigma) = False$. Otherwise,

$$\delta'(\langle v, q \rangle, \sigma) = \bigwedge_{\langle v', q' \rangle \in S_{v,q}^\sigma} (\tau_O(v'), \langle v', q' \rangle).$$

$\square$

Since the UCT $\mathcal{A}_\mathcal{G}$ constructed in Theorem 3 is not deterministic in its V-component, its translation to an NBT involves an exponential blow-up in both $|G|$ and $|\mathcal{U}|$ [11]:

Theorem 4. *Let $\mathcal{G} = \langle G, \mathcal{U} \rangle$ be a game with longitudinal observability, where G is a game graph and $\mathcal{U}$ is a UCW. We can construct an NBT $\mathcal{A}_\mathcal{G}'$ over $(2^O \cup \{\oslash\})$-labeled 2^O-trees such that there is a winning strategy for PLAYER 1 in $\mathcal{G}$ iff $L(\mathcal{A}_\mathcal{G}')$ is not empty. The size of $\mathcal{A}_\mathcal{G}'$ is exponential in $|\mathcal{U}|$ and $|G|$.*

We can now conclude with the following upper bound:

Corollary 2. *Deciding whether PLAYER 1 wins in a longitudinal-observability game $\langle G, \mathcal{U} \rangle$ for a UCW $\mathcal{U}$ is in EXPTIME. The problem can be solved in time exponential in both $|\mathcal{U}|$ and $|G|$. Moreover, when PLAYER 1 wins, the algorithm returns a witness strategy by means of a transducer of size exponential in $|\mathcal{U}|$ and $|G|$.*

3.3   Games with Transverse Observability

For games with transverse observability, we need the following notations. For a vertex $v \in V_2$, a $(V_2^+ \cdot V_1)$-*path from* v is a finite path $v_1, v_2, \ldots, v_k \in V_2^+ \cdot V_1$ in G such that $v_1 = v$. Likewise, a V_2^ω-*path from* v is an infinite path $v_1, v_2, \ldots \in V_2^\omega$ in G such that $v_1 = v$. When PLAYER 1 moves the token to a vertex $v \in V_2$, the token may traverse a $(V_2^+ \cdot V_1)$-path ρ from v, in which case it return to V_1 in $\mathsf{Last}(\rho)$, or it may traverse a V_2^ω-path from v, in which case it never returns to a vertex in V_1.

Consider a UCW $\mathcal{U} = \langle 2^{AP}, Q, q_0, \delta, \alpha \rangle$. Reasoning about games with transverse observability, we are sometimes interested in reachability via a nonempty path that visits α. For this, we define $\delta_\alpha : 2^Q \times \Sigma^+ \to 2^Q$ as follows. First, $\delta_\alpha(S, \sigma) = \delta(S, \sigma) \cap \alpha$. Then, for a word $w \in \Sigma^+$, we define $\delta_\alpha(S, w \cdot \sigma) = \delta(\delta_\alpha(S, w), \sigma) \cup (\delta(S, w \cdot \sigma) \cap \alpha)$. Thus, either α is visited in the prefix of the run that reads w after leaving S, or the last state of the run is in α. It is not hard to prove by an induction on the length of w that for all states $q \in Q$, we have that $q \in \delta_\alpha(S, w)$ iff there is a run from S on w that reaches q and visits α after leaving S.

Assume now that the token is placed in some vertex $v \in V_1$ and that the objective of PLAYER 1 is to force the token into computations in $L(\mathcal{U}^q)$. Assume further that PLAYER 1 chooses to move the token to a successor v' of v. We distinguish between two possibilities.

- $v' \in V_1$. Then, the new objective of PLAYER 1 is to force the token from v' into computations in $L(\mathcal{U}^{q'})$, for all the states $q' \in \delta(q, \tau(v))$.
- $v' \in V_2$. Then, we distinguish between two cases.
  - There is a V_2^ω-path ρ from v' and $\tau(\rho) \notin L(\mathcal{U}^{q'})$ for some $q' \in \delta(q, \tau(v))$. We then say that v' *is a trap for* $\langle v, q \rangle$. Indeed, PLAYER 2 can stay in vertices in V_2 and force the token into a computation not in $L(\mathcal{U}^{q'})$.
  - v' is not a trap for $\langle v, q \rangle$, in which case, for every $(V_2^+ \cdot V_1)$-path $\rho \cdot v''$ from v', PLAYER 1 should force a token that is placed in v'' into computations in $L(\mathcal{U}^{q'})$, for all states $q' \in \delta(q, \tau(v) \cdot \tau(\rho))$.

The above intuition motivates the following definition of *updated objectives*. For a pair $\langle v, q \rangle \in V_1 \times Q$, standing for an objective of PLAYER 1 to force a token placed on v to be accepted from q, and a choice $v' \in V$ of a successor of v, we define the set $S_{v,q}^{v'} \subseteq (V \times Q \times \{\bot, \top\}) \cup \{\mathit{False}\}$ of updated objectives – these that PLAYER 1 has to satisfy in order to fulfil her $\langle v, q \rangle$ objective after choosing v'. The $\{\bot, \top\}$ flag in the updated objectives is used for tracking visits in α: an updated objective $\langle v'', q', c \rangle \in S_{v,q}^{v'}$ has $c = \top$ if PLAYER 2 can force a visit in α when $\mathcal{U}$ runs from q to q' along a word that labels a path from v via v' to v''. Formally, we define $S_{v,q}^{v'}$ as follows. First, if v' is a trap for $\langle v, q \rangle$, then $S_{v,q}^{v'} = \{\mathit{False}\}$. Indeed, once PLAYER 1 chooses a vertex that is a trap for $\langle v, q \rangle$, she cannot fulfil her objective. Otherwise, $S_{v,q}^{v'} \subseteq (V \times Q \times \{\bot, \top\})$, and a triple $\langle v'', q', c \rangle$ is in $S_{v,q}^{v'}$ iff one the following holds.

- $v' \in V_1$, $v'' = v'$, and $q' \in \delta(q, \tau(v))$. Then, $c = \top$ iff $q' \in \alpha$.

- $v' \in V_2$, there is a $(V_2^+ \cdot V_1)$-path $\rho \cdot v''$ from v', and $q' \in \delta(q, \tau(v) \cdot \tau(\rho))$. Then, $c = \top$ iff there is a $(V_2^+ \cdot V_1)$-path $\rho \cdot v''$ from v' such that $q' \in \delta_\alpha(q, \tau(v) \cdot \tau(\rho))$.

Note that it may be that if v' is not a trap for $\langle v, q \rangle$, yet there is no $(V_2^+ \cdot V_1)$-path from v'. That is, all the paths from v' stay in vertices in V_2 and are in $L(\mathcal{U}^{q'})$ for all $q' \in \delta(q, \tau(v))$. Then, $S_{v,q}^{v'} = \emptyset$.

Finally, in games with transverse observability, strategies are V-labeled V_1-trees. In more details, a strategy for PLAYER 1 is $f_1 : V_1^* \to V_1$ such that $f_1(\epsilon) = v_0$, and for all $\rho \in \mathsf{PPrefs}_1(G)$, we have that $\langle \mathsf{Last}(\rho), f_1(\rho) \rangle \in E$.

Theorem 5. *Let* $\mathcal{G} = \langle G, \mathcal{U} \rangle$ *be a game with transverse observability, where* G *is a game graph and* $\mathcal{U}$ *is a UCW. We can construct a UCT* $\mathcal{A}_\mathcal{G}$ *over* V-labeled V_1-trees such that $\mathcal{A}_\mathcal{G}$ accepts a V-labeled V_1-tree $\langle V_1^*, \eta \rangle$ iff $\langle V_1^*, \eta \rangle$ is a winning strategy for PLAYER 1. The size of $\mathcal{A}_\mathcal{G}$ is polynomial in $|G|$ and $|\mathcal{U}|$.

Proof. Let $\mathcal{U} = \langle 2^{AP}, Q, q_0, \delta, \alpha \rangle$. We define $\mathcal{A}_\mathcal{G} = \langle V, V_1, Q', q_0', \delta', \alpha' \rangle$, where

- $Q' = V \times Q \times \{\bot, \top\}$. Intuitively, when $\mathcal{A}$ is in state $\langle v, q, c \rangle$, it accepts strategies that force a token placed on v into a computation accepted by $\mathcal{U}^q$.
- $q_0' = \langle v_0, q_0, \bot \rangle$
- For all $\langle v, q, b \rangle \in V \times Q \times \{\bot, \top\}$ and letter $v' \in V$, if $S_{v,q}^{v'} = \{False\}$ or $\neg E(v, v')$, then $\delta'(\langle v, q, b \rangle, v') = False$. Otherwise,

$$\delta'(\langle v, q, b \rangle, v') = \bigwedge_{\langle v'', q', c \rangle \in S_{v,q}^{v'}} (v'', \langle v'', q', c \rangle).$$

Thus, for every updated requirement $\langle v'', q', c \rangle \in S_{v,q}^{v'}$, the automaton sends a copy in state $\langle v'', q', c \rangle$ to direction v''. Note that several updated requirements may be sent to the same direction. Indeed, different $(V_2^+ \cdot V_1)$-paths from v' may induce different words that $\mathcal{U}$ reads from q. Moreover, since $\mathcal{U}$ is universal, it may send copies in different states even for a single word. Note, however, that all copies sent to direction v'' agree on their V-component, which is v''. Note also that when $S_{v,q}^{v'} = \emptyset$, we get that $\delta'(\langle v, q, b \rangle, v') = True$.

- $\alpha' = V \times Q \times \{\top\}$. Recall that a $\top$ flag indicates that PLAYER 2 may reach the Q-component in an updated requirement traversing a path that visits α. Accordingly, the co-Büchi requirement to visit α only finitely many times amounts to a requirement to visit states with $\top$ only finitely many times.

$\square$

As in the case of games with full observability, the fact $\mathcal{A}_G$ is deterministic in the V-component implies we can translate $\mathcal{A}_\mathcal{G}$ to an NBW $\mathcal{A}_\mathcal{G}'$ that is exponential in $|\mathcal{U}|$ and only polynomial in $|G|$, leading to the following upper bound:

Corollary 3. *Deciding whether* PLAYER 1 *wins in a transverse-observability game* $\langle G, \mathcal{U} \rangle$ *for a UCW* $\mathcal{U}$ *is in EXPTIME. The problem can be solved in time polynomial in* $|G|$ *and exponential in* $|\mathcal{U}|$. *Moreover, when* PLAYER 1 *wins, the algorithm returns a witness strategy by means of a transducer of size polynomial in* $|G|$ *and exponential in* $|\mathcal{U}|$.

4   The Overall Picture

In the upper bounds described in Sect. 3, we assume that the winning condition is given by a UCW. In Fig. 1 below, we describe bounds for the cases the winning condition is given by a UPW, LTL formula, or a DPW.

	full	longitudinal	transverse
UPW	$poly(|G|)$ $exp(|\mathcal{U}|)$	$exp(|G|)$ $exp(|\mathcal{U}|)$	$poly(|G|)$ $exp(|\mathcal{U}|)$
LTL	$poly(|G|)$ $2exp(|\psi|)$	$exp(|G|)$ $2exp(|\psi|)$	$poly(|G|)$ $2exp(|\psi|)$
DBW	$poly(|G|)$ $poly(|\mathcal{D}|)$	$exp(|G|)$ $poly(|\mathcal{D}|)$	$poly(|G|)$ $exp(|\mathcal{D}|)$

Fig. 1. Tight complexity in both parameters in different settings.

The complexities for UPWs coincide with these for UCW. In the full and longitudinal settings, the algorithm is similar to the one in the UCW case. In the transverse setting, the $\{\bot, \top\}$ flag is replaced by a $\{1, \ldots, k\}$ flag, for the index k of the UPW, maintaining the minimal color that is visited during the hidden sub-play (see details in [9]).

Since an LTL formula can be translated to a UCW with an exponential blow-up (say, by dualizing the exponential translation to NBWs [18]), the upper bounds for LTL follow immediately from these for UCW.

Finally, the case of DBW is of special interest, as it is the first to show that the full-observability setting is easier than the transverse-observability setting. Recall the construction of the UCT $\mathcal{A}_\mathcal{G}$ in Theorems 1, 3, and 5. If instead of the UCW $\mathcal{U}$ for the winning condition, we start with a DBW $\mathcal{D}$, then in Theorem 1, the obtained automaton $\mathcal{A}_\mathcal{G}$ is an NBT of size polynomial in both $|\mathcal{G}|$ and $|\mathcal{D}|$. Indeed, from state $\langle v, q \rangle$, the automaton sends to each direction $v' \in V$ at most one updated objective $\langle v', q' \rangle$. Then, in Theorem 3, the obtained UCW is deterministic in its Q-component: from state $\langle v, q \rangle$, the updated objectives sent to a direction $\theta' \in 2^O$ agree on their Q-component, which is $\delta(q, \tau(v))$. Finally, in Theorem 5, the determinism of $\mathcal{D}$ is not helpful, as updated objectives that are sent to the same direction may differ in their Q-component due to different words that $\mathcal{D}$ reads. Thus, in the case of full observability the problem is polynomial in both parameters, in longitudinal observability it is polynomial in the winning condition and exponential in the game graph, and the transverse observability it is exponential in the winning condition and polynomial in the game graph.

We note that all the bounds in the table are tight, and this refers also to the complexities in the different parameters. For example, deciding games with longitudinal observability is EXPTIME-hard even for winning conditions of a fixed size [3], and deciding games with transverse observability is EXPTIME-hard even for games of a fixed size [9].

Finally, the longitudinal and transverse settings share some interesting theoretical properties: both are not determined (that is, there may be games in which no player has a winning strategy), and in both the observability type of PLAYER 2 is not important, in the sense that PLAYER 1 wins against an opponent with partial observability iff she wins against an opponent with full observability. Intuitively, this follows from the fact that in order for a strategy of PLAYER 1 to be winning, it has to win the games against all strategies of PLAYER 2, including the ones that happen to behave as if PLAYER 2 has full observability.

Interesting directions for future research concern settings that combine the two types of partial observability [8], and settings in which the objectives of the players are not contradicting, thus the game is not zero-sum and one cares about stable outcomes [2,7].

References

1. Agarwal, S., Kodialam, M.S., Lakshman, T.V.: Traffic engineering in software defined networks. In: Proceedings of the 32nd IEEE International Conference on Computer Communications, pp. 2211–2219 (2013)
2. Almagor, S., Avni, G., Kupferman, O.: Repairing multi-player games. In: Proceedings of the 26th International Conference on Concurrency Theory. LIPIcs, vol. 42, pp. 325–339 (2015)
3. Alur, R., Henzinger, T.A., Kupferman, O.: Alternating-time temporal logic. J. ACM **49**(5), 672–713 (2002)
4. Chatterjee, K., Doyen, L.: The complexity of partial-observation parity games. In: Fermüller, C.G., Voronkov, A. (eds.) LPAR 2010. LNCS, vol. 6397, pp. 1–14. Springer, Heidelberg (2010). https://doi.org/10.1007/978-3-642-16242-8_1
5. Chatterjee, K., Doyen, L., Henzinger, T.A., Raskin, J.-F.: Algorithms for omega-regular games with imperfect information. In: Ésik, Z. (ed.) CSL 2006. LNCS, vol. 4207, pp. 287–302. Springer, Heidelberg (2006). https://doi.org/10.1007/11874683_19
6. Fisman, D., Kupferman, O.: Reasoning about finite-state switched systems. In: Namjoshi, K., Zeller, A., Ziv, A. (eds.) HVC 2009. LNCS, vol. 6405, pp. 71–86. Springer, Heidelberg (2011). https://doi.org/10.1007/978-3-642-19237-1_10
7. Gutierrez, J., Perelli, G., Wooldridge, M.J.: Imperfect information in reactive modules games. Inf. Comput. **261**, 650–675 (2018)
8. Kupferman, O., Shenwald, N.: Perspective games with notifications. Submitted (2020)
9. Kupferman, O., Vardi, G.: Perspective games. In: Proceedings of the 34th IEEE Symposium on Logic in Computer Science, pp. 1–13 (2019)
10. Kupferman, O., Vardi, M.Y.: Synthesis with incomplete information. In: Advances in Temporal Logic, pp. 109–127. Kluwer Academic Publishers (2000)
11. Kupferman, O., Vardi, M.Y.: Safraless decision procedures. In: Proceedings of the 46th IEEE Symposium on Foundations of Computer Science, pp. 531–540 (2005)
12. Liberzon, D.: Switching in Systems and Control. Birkhauser, Boston (2003). https://doi.org/10.1007/978-1-4612-0017-8
13. Lustig, Y., Vardi, M.Y.: Synthesis from component libraries. Softw. Tools Technol. Transf. **15**(5–6), 603–618 (2013)

14. Margaliot, M.: Stability analysis of switched systems using variational principles: an introduction. Automatica **42**(12), 2059–2077 (2006)
15. Pnueli, A., Rosner, R.: On the synthesis of an asynchronous reactive module. In: Ausiello, G., Dezani-Ciancaglini, M., Della Rocca, S.R. (eds.) ICALP 1989. LNCS, vol. 372, pp. 652–671. Springer, Heidelberg (1989). https://doi.org/10.1007/BFb0035790
16. Puchala, B.: Asynchronous omega-regular games with partial information. In: Hliněný, P., Kučera, A. (eds.) MFCS 2010. LNCS, vol. 6281, pp. 592–603. Springer, Heidelberg (2010). https://doi.org/10.1007/978-3-642-15155-2_52
17. Reif, J.H.: The complexity of two-player games of incomplete information. J. Comput. Syst. Sci. **29**, 274–301 (1984)
18. Vardi, M.Y., Wolper, P.: Yet another process logic. In: Clarke, E., Kozen, D. (eds.) Logic of Programs 1983. LNCS, vol. 164, pp. 501–512. Springer, Heidelberg (1984). https://doi.org/10.1007/3-540-12896-4_383
19. Vardi, M.Y., Wolper, P.: Automata-theoretic techniques for modal logics of programs. J. Comput. Syst. Sci. **32**(2), 182–221 (1986)

Regular Papers

Reachability Set Generation Using Hybrid Relation Compatible Saturation

Shruti Biswal$^{(\boxtimes)}$ and Andrew S. Miner

Iowa State University, Ames, IA 50011, USA
{sbiswal,asminer}@iastate.edu

Abstract. Generating the state space of any finite discrete-state system using symbolic algorithms like saturation requires the use of decision diagrams or compatible structures for encoding its reachability set and transition relations. For systems that can be formally expressed using ordinary Petri Nets (PN), implicit relations, a static alternative to decision diagram-based representation of transition relations, can significantly improve the performance of saturation. However, in practice, some systems require more general models, such as self-modifying Petri nets, which cannot currently utilize implicit relations and thus use decision diagrams that are repeatedly rebuilt to accommodate the changing bounds of the system variables, potentially leading to overhead in saturation algorithm. This work introduces a hybrid representation for transition relations, that combines decision diagrams and implicit relations, to reduce the rebuilding overheads of the saturation algorithm for a general class of models. Experiments on several benchmark models across different tools demonstrate the efficiency of this representation.

Keywords: Petri Nets · Self-modifying nets · Decision diagrams · Saturation · Reachability set generation · Implicit relations

1 Introduction

With rapid progress in the development of automated systems, formal verification techniques have increasingly aided in building thorough and reliable automation. One such technique, model checking, involves an exhaustive analysis of all possible systemic behavior, conventionally defined in terms of its variables and events, for a design to be verified. This motivates the need for efficient state-space generation methods, in spite of the fact that the reachability set of a system can be extremely large due to state explosion problem.

State-of-the-art symbolic techniques for state-space generation, such as *saturation* [4], are time- and memory-efficient with respect to the explicit methods. Several implementations of the saturation algorithm demonstrate that the technique has been often fine-tuned to perform efficiently with different sub-classes of discrete-state systems. The most generic implementation of the saturation algorithm utilizes multi-valued decision-diagrams (MDDs) [13] as the underlying data structure for representation of reachable states and extensible matrix

© Springer Nature Switzerland AG 2020
S. Schmitz and I. Potapov (Eds.): RP 2020, LNCS 12448, pp. 37–51, 2020.
https://doi.org/10.1007/978-3-030-61739-4_3

diagrams (MxDs) [4, 20] for representation of system events where the bounds of the states variables are not known a priori. For systems that can be represented in terms of Kronecker products, a static encoding for the events, such as *implicit relation forests* [3], is an efficient alternative that aids saturation by eliminating the need to rebuild transition relations as state variable sizes increase, thereby allowing the algorithm to focus completely on generation of reachability set.

In the real world, however, systems may have some events that are not expressible as Kronecker products, or are expressible as Kronecker products except for a few components. Such models would require either to use matrix diagrams entirely, or to split events until each event is expressible as a Kronecker product so that implicit relation forests may be used. The aim of the paper is to explore a *hybrid relation*, that allows the mixture of implicit nodes (from implicit relation forests) and matrix diagram nodes in the same forest. This makes it possible to simultaneously use implicit nodes for (portions of) events expressible as a Kronecker product, and matrix diagram nodes for the more general case.

The remainder of the paper is organized as follows. Section 2 describes the formalism for the class of models befitting the proposed methodology, the saturation algorithm and briefly discusses the various saturation-compatible data structures available in the literature. Section 3 introduces hybrid relation forests, describes their construction from a model, and formulates a modified saturation algorithm that works with events represented in a hybrid relation forest. Section 4 evaluates the proposed method experimentally by comparing its performance with existing techniques in the literature. Finally, Sect. 5 concludes the paper and presents some directions for future work.

2 Background and Related Work

In order to describe the problem domain, we use a standard definition of extended Petri nets, which are quite general and include inhibitor arcs and marking-dependent arc cardinalities, as a high-level formalism to define the class of finite discrete-state models.

Definition 1 (Extended Petri Net). The high-level formalism of a model $\mathcal{M}$ as an extended Petri net is defined as a tuple $(\mathcal{P}, \mathcal{T}, \mathcal{F}^-, \mathcal{F}^+, \mathcal{F}^o, \mathbf{i}_0)$ where:

- $\mathcal{P} = \{p_1, p_2, \ldots, p_{|\mathcal{P}|}\}$ is a finite set of *places* that encode the state variables of $\mathcal{M}$. Each place can contain a natural number of tokens. A marking $\mathbf{i} = (i_1, i_2, \ldots, i_{|\mathcal{P}|}) \in \mathbb{N}^{|\mathcal{P}|}$ represents the number of tokens in each place.
- $\mathcal{T} = \{t_1, t_2, \ldots, t_{|\mathcal{T}|}\}$ is a finite set of *transitions* of the net that represent the events of $\mathcal{M}$.
- $\mathcal{F}^- : \mathcal{P} \times \mathcal{T} \times \mathbb{N}^{|\mathcal{P}|} \to \mathbb{N}$, $\mathcal{F}^+ : \mathcal{T} \times \mathcal{P} \times \mathbb{N}^{|\mathcal{P}|} \to \mathbb{N}$, and $\mathcal{F}^o : \mathcal{P} \times \mathcal{T} \times \mathbb{N}^{|\mathcal{P}|} \to \mathbb{N} \cup \{\infty\}$ are the marking-dependent input, output, and inhibitor arcs, respectively.
- $\mathbf{i}_0 \in \mathbb{N}^{|\mathcal{P}|}$ is the initial marking of the net.

A transition t is *enabled* on a marking $\mathbf{i}$ iff $\forall p \in \mathcal{P}$, $\mathcal{F}^-(p,t,\mathbf{i}) \leq i_p < \mathcal{F}^o(p,t,\mathbf{i})$. When multiple transitions are enabled on a marking, the choice for *firing* a transition is made non-deterministically. On firing an enabled transition t on $\mathbf{i}$, a new marking $\mathbf{j}$ is obtained, where $\mathbf{j}_p = \mathbf{i}_p - \mathcal{F}^-(p,t,\mathbf{i}) + \mathcal{F}^+(t,p,\mathbf{i})$, $\forall p \in \mathcal{P}$.

The next-state function $\mathcal{N}_t(\mathcal{I}) = \{\mathbf{j} : \mathbf{i} \in \mathcal{I}$ and $\mathbf{i} \xrightarrow{t} \mathbf{j}\}$ represents the next set of markings obtained when transition t is fired on a given set of markings, $\mathcal{I}$. The support-set of a next-state function, defined as $supp(\mathcal{N}_t) = \{p \in \mathcal{P} : \exists \mathbf{i} \in \mathbb{N}^{|\mathcal{P}|}, \mathcal{F}^-(p,t,\mathbf{i}) > 0 \vee \mathcal{F}^+(t,p,\mathbf{i}) > 0 \vee \mathcal{F}^o(p,t,\mathbf{i}) < \infty\}$, is the set of places that could affect the enabling of t, or could affect or be affected by the firing of t.

Given the initial marking $\mathbf{i}_0$, a marking $\mathbf{i}'$ is said to be reachable if there exists a finite sequence of transitions which when fired in succession starting from $\mathbf{i}_0$ can lead to $\mathbf{i}'$. The set of all possible reachable markings of a Petri net, denoted by $\mathcal{S} \subseteq \mathbb{N}^{|\mathcal{P}|}$, is the least fixed point such that $\mathcal{S} = \{\mathbf{i}_0\} \cup \bigcup_{t \in \mathcal{T}} \mathcal{N}_t(\mathcal{S})$.

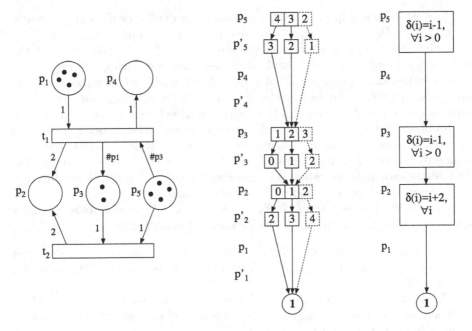

Fig. 1. A Petri net model with marking-dependent arcs, with 2L-level MDD and implicit relation representation of transition t_2.

The objective of this work is to generate the set $\mathcal{S}$ using saturation-based symbolic state-space generation. The algorithms in the paper are based on the assumption that $\mathcal{S}$ is finite. Therefore, all state variables of the model are bounded, but these bounds do not need to be known a priori as they are discovered during execution of the algorithm.

2.1 Symbolic Representation of State Space and Transition Relations

The state-space of a model is a set of all reachable states, represented symbolically as an indicator function encoded using decision diagrams [19].

Definition 2 (Multi-valued Decision Diagram [13]**).** Defined over an ordered sequence of L domain variables $(v_L, \ldots, v_1)$ such that $v_{k+1} \succ v_k$ and the variable-specific domains are denoted as $\mathcal{D}_k = \{0, 1, \ldots, n_k\}$, an ordered MDD is a directed acyclic graph where:

- *Terminal* nodes are labeled **0** or **1** and are associated with a special variable v_0 such that any domain variable $v_k \succ v_0$.
- Every *non-terminal* node p is associated with some domain variable p.var = $v_k \succ v_0$ and has $\mathcal{D}_k$ child pointers denoted by p$[i_k]$ such that p.$var \succ$ p$[i_k]$.var.

Two nodes, p and q, in an MDD are said to be *duplicates* if p.var = q.var = v_k and $\forall i_k \in \mathcal{D}_k$, p$[i_k]$ = q$[i_k]$. If all pointers of a node point to the same child, then the parent node is termed as *redundant node*. To correctly interpret the unknown but finite variable bounds which are explored during generation, *on-the-fly* saturation requires the use of *quasi-reduced* MDDs, which forbid duplicate nodes but require redundant nodes except for all **0** children: for every non-terminal node p and for every child i_k, either p$[i_k]$ = **0** or p$[i_k]$.var = p.var − 1.

Similar to MDDs, a transition relation of a model can be encoded in the form of a set of pairs of states and their next-states using 2L-level MDDs where an indicator function is defined over $(\mathcal{D}_k \times \mathcal{D}_k) \times \ldots \times (\mathcal{D}_1 \times \mathcal{D}_1) \to \{0, 1\}$. The first set in each pair corresponds to unprimed or "from" state of variable and the second set refers to primed or "to" state of variable.

Given a Petri net encoding $(\mathcal{P}, \mathcal{T}, \mathcal{F}^-, \mathcal{F}^+, \mathcal{F}^o, \mathbf{i}_0)$ of a model, its reachability set, $\mathcal{S}$ and transition relation, $\mathcal{N}$ can be encoded using MDDs:

- The domain variables $(v_L, \ldots, v_1)$ of the MDD correspond to the Petri net $\mathcal{P} = (p_1, p_2, \ldots p_{|\mathcal{P}|})$ such that each variable v_k is associated with one p_i, for simplicity.
- The next-state function, $\mathcal{N}_t(\mathbf{i}) = \mathbf{j}$ can be represented by a 2L-level MDD where nodes corresponding to unprimed level and primed level of variable v_k encode the pair (i_k, j_k).

Nodes in a 2L-level MDD can be redundant or duplicate under same definitions as above. While a 2L-level MDD can be *reduced* as per any choice of reduction rule, the proposed work requires the use of *Fully-Fully* [7] reduced 2L-level MDD which results in skipping of redundant nodes.

Figure 1 shows a small Petri net (left), and symbolic encodings of the relation for transition t_2 using a both a 2L-level MDD (middle) and an implicit relation (right). It is noticed that for a Petri net transition where the modified value of a state variable is a function of the variable itself, the 2L-level MDD of the transition assumes a peculiar shape, wherein every edge from the nodes associated with some primed variable v'_k points to the same child node. This structure creates an unnecessary overhead for on-the-fly saturation as it requires to be rebuilt each time the variable bound expands. In order to encode $\mathcal{N}_t$ of such PNs efficiently, [3] proposes an alternate representation, *implicit relation*. The full encoding of $\mathcal{N}$ contains $|\mathcal{T}|$ implicit relations uniquely identified by their topmost nodes, forming an *implicit relation forest*.

Definition 3 (Implicit Relation). An implicit relation defined over the sequence of L domain variables $(v_L, \ldots, v_1)$ is a compact version of a 2L-level MDD where the unprimed and primed levels of a variable are combined into a single node, *relation node* which encodes the local effect of the transition on that variable. Each relation node r has a single downward pointer $r.ptr$ to another node such that $r.var \succ r.ptr.var$.

Previous works of Couvreur et al. [9] have proposed *homomorphisms* as an alternative approach to symbolically encode transition relations that works well with Data Decision Diagrams (DDD) [9] and Hierarchical Set Decision Diagrams (SDD) [11] for reachability generation. Homomorphisms are a class of operators which are used on DDDs to encode the function represented by the transition relation. Additionally, this representation is generic and is designed well [10] for extended Petri nets like 2L-level MDDs. Molnar et al. in [14] proposed *abstract next-state diagrams* that encode the transition relations via the use of level-wise *descriptors* to capture the local effect of the transitions. While the domain of models and the key idea of abstracting Petri net transitions is similar to that of implicit relations, the facility with implicit relations to skip the encoding for non-support variables of transitions provides a computational advantage during saturation. The various symbolic representations of transition relations discussed in this section are used as benchmark techniques for performance comparison of the proposed work.

2.2 Saturation

Saturation is a symbolic algorithm for state-space generation that uses DAG-type data structures like decision diagrams or implicit relations for storing and computing operations on the data. In order to calculate the state-space of any system, a breadth first iteration on the fixed point equation $\mathcal{S} = \{\mathbf{i}_0\} \cup \mathcal{N}(\mathcal{S})$, where $\mathcal{N}(\mathcal{S}) = \bigcup_{t \in \mathcal{T}} \mathcal{N}_t(\mathcal{S})$ is an explicit method and is computationally inefficient. However, the strategy of *node saturation* computes the fixed point recursively by exploring reachable states through continuous computation of *relational product* between $\mathcal{S}$ and $\mathcal{N}$ in a bottom-up fashion. The algorithm is described in procedure Saturate of Fig. 4. On-the-fly saturation [5,15,16] is a variation that explores the variable bounds during reachability set generation and eliminates the need for knowing the variable bounds a priori. However, this dynamic expansion of variable bounds involves certain overhead due to re-construction of $\mathcal{N}$. *Extensible decision diagrams* [20] were introduced to alleviate some of these overheads.

3 Hybrid Relations

Taking into consideration the generality of real-world discrete-state systems and the need to encode these models for efficient execution of on-the-fly saturation techniques, this section formulates an ensemble of decision diagrams and implicit relations for representation of next-state functions where the existence of dependency among variables in the functions and the choice of partitioning technique for the transition relations determine the data structure to be used. This representation combines the power of generalization from 2L-level MDDs and the efficiency of static description from implicit relations. We also modify the on-the-fly saturation algorithm to work with the hybrid relations.

Definition 4 (Hybrid relation). Given a transition $t \in \mathcal{T}$ of a Petri net with a conjunctively-partitioned next-state function $\mathcal{N}_t = \bigcap_{i=1}^{m_t} \mathcal{N}_{t,c_i}$, a hybrid relation encoding of t has a corresponding set of m_t DAGs as follows.

- Each DAG encodes one of the sub-relations $\mathcal{N}_{t,c}$, using either an implicit relation or a 2L-level MDD. Each DAG is uniquely identified by its top-most node.
- Every 2L-level MDD obeys FF-reduction rule for skipping nodes associated with variables $v \notin supp(\mathcal{N}_{t,c})$.
- Every implicit relation contains a single non-terminal relation node with pointer to the terminal $\mathbf{1}$.
- The interpretation of skipped levels is based on the overall hybrid relation of the transition: if a skipped variable v affects the transition at all, then $v \in supp(\mathcal{N}_t)$, and thus $v \in supp(\mathcal{N}_{t,c})$ for some c. In this case, the FF-reduction ensures that the intersection is handled appropriately. If instead v does not affect the transition ad $v \notin supp(\mathcal{N}_t)$, then $v \notin supp(\mathcal{N}_{t,c})$ for any c and the interpretation is "v is not changed by the firing of t".

3.1 Constructing Next-State Functions

We construct the DAG for each sub-relation $\mathcal{N}_{t,c}$ based on the number of variables in its support. The case $supp(\mathcal{N}_{t,c}) = \emptyset$ corresponds to state variables that do not affect, and are not affected by, the firing of t in the sub-relation; we assume that this case is already handled (because each such variable v will either not affect t at all because $v \notin supp(\mathcal{N}_t)$, or will affect t but then $v \in supp(\mathcal{N}_{t,c'})$ for some other c'). When $supp(\mathcal{N}_{t,c}) = \{v\}$, $\mathcal{N}_{t,c}$ can be represented using an implicit relation node r where $r.var = v$, the partial function $r.\delta$ encodes the sub-relation and $r.ptr = \mathbf{1}$. In practice, the implicit relation nodes can be compacted (conjuncted) to form a single implicit relation, which will be the original implicit nodes chained together. This has a performance advantage since it decreases the number of sub-relations that needs to be kept track of during saturation for their *symbolic conjunction*. When $|supp(\mathcal{N}_{t,c})| > 1$, $\mathcal{N}_{t,c}$ must be encoded using a 2L-level MDD over the domain variables in $supp(\mathcal{N}_{t,c})$.

Different methods may be used for partitioning a transition relation $\mathcal{N}_t$ into its sub-relations. One method [16] is to use the finest possible partition, such that each state variable belongs to the support of at most one sub-relation. Another method [7] separates the transition enabling and firing conditions, and then uses the finest possible partition so that each variable is *modified* in at most one sub-relation (but may appear, read-only, in other sub-relations). These methods are illustrated in Fig. 2 for transition t_1 from the PN in Fig. 1. The left side uses the method from [16], and obtains two sub-relations: one as a chain of implicit nodes, representing $p'_4 = p_4 + 1$ and $p'_2 = p_2 + 2$, and one as an MDD, representing $(p_5 \geq p_3) \wedge (p'_5 = p_5 - p_3) \wedge (p'_3 = p_3 + p_1) \wedge (p_1 \geq 1) \wedge (p'_1 = p_1 - 1)$. The right side uses the method from [7], and obtains three sub-relations: one representing $p'_3 = p_3 + p_1$, one representing $(p_5 \geq p_3) \wedge (p'_5 = p_5 - p_3)$, and a chain of implicit nodes representing $p'_4 = p_4 + 1$, $p'_2 = p_2 + 2$, and $(p_1 \geq 1) \wedge (p'_1 = p_1 - 1)$.

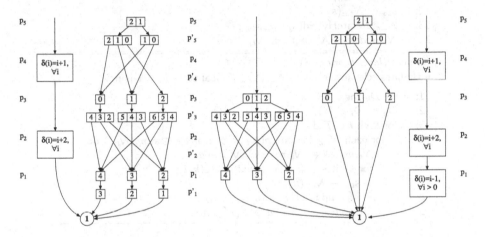

Fig. 2. Hybrid relation-based encodings of transition t_1 from Fig. 1 when partitioning techniques from [16] (left) and [7] (right) are used.

Algorithm BuildHybrid, shown in Fig. 3, constructs an overall hybrid relation for a given transition $t \in \mathcal{T}$, as a set of DAGs. This should be called once, for each transition, before saturation is invoked. The set of sub-relation DAGs for transition t is denoted by $\mathcal{N}_t.nodes$. Note that the overall transition relation $\mathcal{N}_t$ is not computed, but rather is evaluated during the saturation algorithm on the fly. The hybrid relations are updated during saturation as new local states are discovered, using algorithm Confirm, also shown in Fig. 3.

3.2   Saturation Using Hybrid Relations

Saturating a node for variable v_k requires repeatedly firing transitions in $\mathcal{T}_k$, the set of transitions whose top-most variable is v_k (i.e., $supp(\mathcal{N}_t) \cap \{v_L, \ldots, v_k\} = \{v_k\}$), until a fixed point is reached. The algorithm is shown in Fig. 4. Procedures RecFire and Saturate are similar to those of saturation using MxDs [16] (using either a single DAG per transition, or a single DAG for all of $\mathcal{T}_k$), except all nodes

---

BuildHybrid(*transition t*)
• Builds hybrid relation for Petri net transition t

1: Partition $\mathcal{N}_t$ into $\mathcal{N}_{t,c_1}, \mathcal{N}_{t,c_2}, \ldots \mathcal{N}_{t,c_{m_t}}$
2: $n \leftarrow 1$;
3: for each $\mathcal{N}_{t,c}$ s.t. $|supp(\mathcal{N}_{t,c})| = 1$, from bottom up, do
4:     $r \leftarrow$ new relation node;
5:     $r.var \leftarrow \mathcal{N}_{t,c}.var$;
6:     $r.\delta \leftarrow \mathcal{N}_{t,c}.\Delta$;
7:     $r.ptr \leftarrow n$;
8:     $n \leftarrow r.ptr$;
9: $\mathcal{N}_t.nodes \leftarrow \{n\} \setminus \{1\}$;
10: for each $\mathcal{N}_{t,c}$ s.t $|supp(\mathcal{N}_{t,c})| > 1$ do
11:     $s_{from} \leftarrow s_0$;
12:     $s_{to} \leftarrow \mathcal{N}_{t,c}(s_{from})$;
13:     $\mathcal{N}_{t,c}.updateHybrid(s_{from}, s_{to})$;
14:     $\mathcal{N}_t.nodes \leftarrow \mathcal{N}_t.nodes \cup \mathcal{N}_{t,c}.node$;

Confirm(*level k, index i*)
• Updates relevant relations $\mathcal{N}_t$, $\mathcal{N}_{t,c}$ if local state $v_k = i$ is new

1: if $i \notin \mathcal{D}_k$ then
2:     $\mathcal{D}_k \leftarrow \mathcal{D}_k \cup \{i\}$;
3:     for each $t \in \mathcal{T}$ do
4:         for each $\mathcal{N}_{t,c}$ s.t. $|supp(\mathcal{N}_{t,c})| > 1$ and $k \in supp(\mathcal{N}_{t,c})$ do
5:             $\mathcal{N}_t.nodes \leftarrow \mathcal{N}_t.nodes \setminus \mathcal{N}_{t,c}.node$;
6:             for each $s_{from} \in \mathcal{N}_{t,c}.buildState(k, i)$;
7:                 $s_{to} \leftarrow \mathcal{N}_{t,c}(s_{from})$;
8:                 $\mathcal{N}_{t,c}.updateHybrid(s_{from}, s_{to})$
9:             $\mathcal{N}_t.nodes \leftarrow \mathcal{N}_t.nodes \cup \{\mathcal{N}_{t,c}.node\}$;

Fig. 3. Algorithms for building a hybrid relation.

of the hybrid relation for a transition are traversed simultaneously. Thus, in addition to the MDD node encoding the set of states, procedure RecFire requires a set of hybrid relation nodes, rather than a single MxD node (as used in [16]) or a transition identifier (as used in [5]). This also means that the compute table entries for RecFire must include the set of hybrid relation nodes (c.f. lines 2 and 15).

The actual firing with a set of hybrid relation nodes is handled in helper procedure Fire, also shown in Fig. 4. This takes in a set of hybrid relation nodes, and determines the set of nodes to follow one level below, by examining each node in the set and following its downward pointer ($h[i][j]$ in line 4), unless the current level is skipped, in which case the current node is used again (line 3). Note that $h[i][j]$ must be determined differently, based on if h is an implicit node, or an MDD node. If the set of downward pointers contains terminal node **0**, then the intersection of the sub-relations will be empty, and in this case no states will be reached (line 5). Otherwise, the traversal continues by calling RecFire recursively

---

mdd Fire(mdd n, nodeset $\mathcal{H}$, index i, j)
• Fire hybrid relation nodes $\mathcal{H}$ on node n, from i to j.

1: $\mathcal{H}' \leftarrow \emptyset$;   • Determine next-level relation nodes
2: for each $h \in \mathcal{H}$ do
3:    if $n.var > h.var$ then $\mathcal{H}' \leftarrow \mathcal{H}' \cup \{h\}$;
4:    else $\mathcal{H}' \leftarrow \mathcal{H}' \cup \{h[i][j]\}$;
5: if $\mathcal{H}' = \emptyset$ or $\mathbf{0} \in \mathcal{H}'$ then return $\mathbf{0}$;
6: $f \leftarrow$ RecFire($n[i], \mathcal{H}'$);
7: if $f \neq \mathbf{0}$ then Confirm($n.var, j$);
8: return f;

Saturate($level$ k, mdd n)
• Saturate node n at level k using $\mathcal{T}_k$.

1: $\mathcal{Q} \leftarrow \{i : n[i] \neq \mathbf{0}\}$;
2: while $\mathcal{Q} \neq \emptyset$ do
3:    $i \leftarrow$ Choose($\mathcal{Q}$);
4:    $\mathcal{Q} \leftarrow \mathcal{Q} \setminus \{i\}$;
5:    for each $t \in \mathcal{T}_k$
6:       $h^t \leftarrow$ topmost($\mathcal{N}_t.nodes$);
7:       for each j s.t. $h^t[i][j] \neq \mathbf{0}$ do
8:          $f \leftarrow$ Fire($n, \mathcal{N}_t.nodes, i, j$);
9:          $u \leftarrow$ Union($f, n[j]$)
10:         if $u \neq n[j]$ then
11:            $n[j] \leftarrow u$;
12:            $\mathcal{Q} \leftarrow \mathcal{Q} \cup \{j\}$;

mdd RecFire(mdd n, nodeset $\mathcal{H}$)
• Fire $\mathcal{H}$ on node n and then saturate

1: if $n = \mathbf{0}$ or $\mathcal{H} = \{\mathbf{1}\}$ then return n;
2: if (RecF, $n, \mathcal{H}, m$) $\in$ CT then return m;
3: $h^t \leftarrow$ topmost($\mathcal{H}$);
4: $k \leftarrow$ max($n.var, h^t.var$);
5: $m \leftarrow$ new MDD node for variable v_k;
6: if $n.var > h^t.var$ then
7:    for each $i \in \mathcal{D}_k$ do
8:       $m[i] \leftarrow$ RecFire($n[i], \mathcal{H}$);
9: else
10:   for each i, j s.t. $n[i] \neq \mathbf{0}, h^t[i][j] \neq \mathbf{0}$
11:      $f \leftarrow$ Fire($n, \mathcal{H}, i, j$);
12:      $m[j] \leftarrow$ Union($m[j], f$);
13: Saturate(k, m);
14: $m \leftarrow$ UniqueInsert(m);
15: CT $\leftarrow$ CT $\cup$ (RecF, $n, \mathcal{H}, m$);
16: return m;

Fig. 4. Saturation using hybrid relations

(line 6). If the reached set of states is non-empty, then local state variable j must be confirmed [5] if it is new (line 7).

4   Experimental Results

4.1   Setup

We implemented the proposed hybrid relation-based saturation algorithm, HybSat, using the [16] partitioning method in SMART [6] using Meddly [2] as the underlying decision diagram library. We conducted two groups of experiments to compare the performance of HybSat,

– **Group I**, with the existing "on-the-fly saturation with matrix diagrams" approach (OtfSat) from SMART and with DDD-based reachability set generation from ITS-Tools (ITSTools) for extended Petri net models on identical static variable order for each run.

– **Group II**, with the saturation algorithm based on a generalized representation of next-state functions (AbstractDesc) from [17] for ordinary Petri net models using SOUPS [18] ordering for variables.

Each Petri net model has multiple instances described by parameters that determine the size of the model or its initial state. Each experimental run comprises of running an instance of a model on the group-specific set of tools. SMART and ITS-Tools are run on their default settings from MCC, except that the use of SDD in ITS-Tools is disabled since it requires additional structural information about the model. The tool for AbstractDesc is Java-based and is allocated a maximum heap size of 6 GB and stack size of 512 MB for its experiments. All experiments are run on a server of Intel Xeon CPU 2.13 GHz with 48G RAM under Linux Kernel 4.9.9 with timeout for each run set to 1 h.

For Group II experiments, we obtained the ordinary Petri net models from the model repository of [1]. The suite of extended Petri nets used for the experiments in Group I category are described as follows:

– *Swap* model where given a list of N distinct integers, operations are to exchange two neighboring integers. The model contains N places and N-1 transitions, where both arcs of each transition have marking-dependent cardinalities.
– *Stack* model that generates the count of possible combination of D objects that can have values in the range from 1 to N. The PN has D places and 4 transitions, where inc & dec control that the number of tokens are within limits of N and push and pop move the tokens among the places of the net.
– A simplified version of *Leader* election protocol [12] that designates a single process as the leader among a ring of N processes by determining the maximum of the *unique_id*s transmitted by the processes.
– A non-stochastic version of *FMS* (Flexible Manufacturing System) from [8] that has N number of 3 different kinds of parts fitted on three machines.
– *Tiles* models a N × M rectangular puzzle with movable tiles. The model has N × M places and 4 transitions (left, right, up, down) to represent position and movement of the tiles.
– *Tower of Hanoi* models D disks of different sizes that can move across three rods with constraints on sizes of the disks that can be placed on a rod.
– A modified version of *SatelliteMemory*, an ordinary Petri net from [1], where the constant arc cardinalities are converted to constant places and markings of these places are used to define the arc cardinalities in order to introduce synthetic marking-dependent behavior.

4.2   Observations

Table 1 and Table 2 summarize the experiments run on extended Petri nets and ordinary Petri nets respectively. The metrics used for performance comparison include the time taken by the reachability generation algorithm and the total number of peak nodes required by the execution for encoding the transition relations. For experiments conducted under Group I (see Table 1), the observations from OtfSat and HybSat reveal that models that have events with tightly-bound sub-event effects, force the HybSat to build 2L-level MDDs leading to similar metric values and often these values are worse with respect to ITSTools. For example, models like Stack, the sub-events display a chain-dependency among the variables leading to creation of a full 2L-level MDD for encoding the relation. For models such as FMS, where all but four transitions have constant arc cardinalities, HybSat performs better than ITSTools but shows similar performance with OtfSat. The overall results in these experiments indicate that higher the number of relation nodes in the encoding, better is the performance of HybSat. However, the ability of HybSat to encode general transitions make it as versatile as OtfSat and ITSTools.

The observations of Group II experiments that are summarized in Table 2, confirm that within a given model, the ratio of execution times of saturation algorithm in the Java implementation of AbstractDesc to that of HybSat in SMART is fairly constant. While these encodings and algorithms are conceptually similar, the significant yet constant difference in the execution times can be attributed to the implementational details and underlying language. Note that the number of nodes in case of AbstractDesc is higher compared to HybSat because the former does not skip descriptors that encode identity relations and it counts every pair of "from" and "to" local states as defined by the descriptors. The number of nodes encoded for $\mathcal{N}$ by HybSat for is at most equal to the number of arcs in the Petri net definition. Since HybSat pre-determines the encoding strategy for each transition, for all models under this experiment group, their transitions are represented as relation nodes. Hence, the comparison metrics for reachability generation in HybSat and ImplSat are exactly the same and therefore, the observations of ImplSat are omitted. The missing fields refer to timeout of the experimental run.

5 Conclusions

In this paper, we introduced hybrid relations, an ensemble data-structure, for symbolic encoding of transition relations in a general class of discrete-state systems. The translation of a system event into its corresponding hybrid relation, involves structural analysis of the event in terms of dependencies among the variables of the event. This prior system review to assign a suitable data-structure for encoding each (sub)event rewards during saturation, since unnecessary overhead costs of managing the compute-table entries for 2L-level MDD operations and saturation are eliminated. The efficiency and validity of the proposed saturation algorithm is empirically evaluated with respect to several versions of saturation algorithm.

Table 1. Performance comparison of reachability generation among SMART (OtfSat, HybSat) and ITSTools from ITS-tools over benchmark of extended Petri net models

Instance	$|\mathcal{S}|$	Time (in sec)			Peak Nodes: $\mathcal{N}$		
		OtfSat	ITSTools	HybSat	OtfSat	ITSTools	HybSat
Swap							
10	3.6×10^6	2.7×10^{-01}	9×10^{-01}	3.3×10^{-01}	1149	58	1091
12	4.8×10^8	3.0×10^{00}	9.5×10^{00}	3.5×10^{00}	1890	70	1861
15	1.3×10^{12}	3.2×10^{02}	4.0×10^{02}	2.9×10^{02}	3360	88	3586
Stack							
D8 V5	3.9×10^5	2.7×10^{00}	6.0×10^{-02}	2.7×10^{00}	290	41	242
D10 V3	5.2×10^{47}	6.3×10^2	5.0×10^{00}	5.7×10^2	2763	409	2120
Leader							
7	2.39×10^7	4.0×10^{01}	5.8×10^{01}	6.1×10^{01}	1109	79	476
8	3.0×10^8	1.9×10^{02}	2.3×10^{02}	2.7×10^{02}	1676	105	735
Flexible manufacturing system							
15	2.7×10^{11}	2.9×10^{-01}	5.3×10^{-01}	1.7×10^{-01}	1544	118	355
20	6.0×10^{12}	4.2×10^{-01}	1.29×10^{00}	3.9×10^{-01}	1660	118	505
30	7.7×10^{14}	9.1×10^{-01}	5.51×10^{00}	1.1×10^{00}	2800	118	880
40	2.6×10^{16}	6.4×10^{00}	1.3×10^{01}	5.3×10^{00}	7342	118	1355
Tiles							
N3M3	1.8×10^{05}	1.4×10^{00}	8.9×10^{00}	2.2×10^{00}	952	154	362
N5M2	1.8×10^{06}	1.5×10^{01}	5.5×10^{01}	1.3×10^{01}	1223	168	433
N3M4	2.4×10^{08}	2.3×10^{01}	3.1×10^{01}	1.7×10^{01}	2561	212	641
Tower of Hanoi							
D8	6.5×10^{03}	3.7×10^{00}	5.9×10^{00}	4.3×10^{00}	4566	40	1219
D10	5.9×10^{04}	5.4×10^{01}	7.0×10^{01}	5.9×10^{01}	9837	40	2481
Satellite							
X 100 Y 3	7.6×10^{04}	2.1×10^{01}	1.1×10^{01}	1.9×10^{01}	3207	98	737

Although this paper briefly discusses the use of event partitioning techniques for building hybrid relations, the consequence of each technique on the efficiency of hybrid relations based saturation algorithm is yet to be explored. This can pave way for applying desirable partitioning methods for each event instead of making an apriori choice for all events. Building and incorporating compatible data structures into hybrid relations that can optimally utilize certain event behaviors can make the representational formalism more powerful and efficient for saturation-based symbolic reachability generation.

Table 2. Performance comparison of reachability generation among SMART (ImpSat, HybSat) and AbstractDesc over benchmark of ordinary Petri net models

Model	$|\mathcal{S}|$	Time (in sec)		Peak Nodes for encoding $\mathcal{N}$	
		AbstractDesc	HybSat	AbstractDesc	HybSat
Eratosthenes					
200	1.1×10^{46}	4.09×10^{-01}	2.27×10^{-02}	441	603
500	4.1×10^{121}	2.32×10^{00}	2.35×10^{-01}	2820	1875
Philosophers					
1000	1.3×10^{477}	2.24×10^{01}	3.98×10^{00}	62410	15000
2000	4.0×10^{2385}	7.03×10^{01}	2.41×10^{01}	124917	30000
SwimmingPool					
10	3.4×10^{10}	5.87×10^{01}	3.97×10^{01}	721877	20
Kanban					
200	3.2×10^{22}	1.26×10^{01}	1.92×10^{00}	44028	39
500	7.1×10^{26}	1.51×10^{02}	2.94×10^{01}	260027	39
HouseConstruction					
20	1.4×10^{13}	1.27×10^{01}	3.06×10^{00}	279043	51
50	1.6×10^{19}	1.74×10^{02}	4.37×10^{01}	2280222	51
SmallOperatingSystem					
MT512 DC128	1.0×10^{11}	1.51×10^{02}	4.05×10^{01}	348039	22
MT512 DC256	2.5×10^{11}	3.13×10^{02}	1.03×10^{02}	594183	22
GPPP					
C10 N10	2.4×10^{10}	7.1×10^{00}	9.99×10^{00}	173153	79
C10 N100	1.8×10^{11}	7.25×10^{00}	1.34×10^{01}	194454	79
Referendum					
100	5.2×10^{47}	4.48×10^{-01}	1.57×10^{-02}	2701	460
200	2.7×10^{95}	3.14×10^{01}	9.07×10^{00}	529255	940
TCPcondis					
15	5.4×10^{12}	3.29×10^{02}	8.48×10^{01}	4519532	90
20	5.6×10^{14}	–	3.03×10^{02}	–	90
Raft					
6	2.9×10^{26}	–	1.87×10^{01}	–	620
7	3.6×10^{35}	–	1.46×10^{02}	–	829
Peterson					
2	2.1×10^{04}	1.79×10^{00}	1.50×10^{01}	18055	298
3	3.4×10^{06}	1.29×10^{02}	2.85×10^{01}	1451193	739
Airplane					
200	2.8×10^{08}	1.21×10^{02}	1.08×10^{01}	1208062	2172
500	4.3×10^{09}	–	4.44×10^{00}	–	6376
Dekker					
15	2.8×10^{05}	7.05×10^{02}	3.99×10^{01}	7084909	729
20	1.2×10^{07}	5.21×10^{02}	1.31×10^{01}	3565360	1365

Acknowledgment. This work was supported in part by the National Science Foundation under grant ACI-1642327.

References

1. MCC: Model Checking Competition @ Petri Nets. https://mcc.lip6.fr
2. Babar, J., Miner, A.S.: MEDDLY: multi-terminal and edge-valued decision diagram LibrarY. In: Proceedings of the QEST, pp. 195–196. IEEE Computer Society (2010)
3. Biswal, S., Miner, A.S.: Improving saturation efficiency with implicit relations. In: Donatelli, S., Haar, S. (eds.) PETRI NETS 2019. LNCS, vol. 11522, pp. 301–320. Springer, Cham (2019). https://doi.org/10.1007/978-3-030-21571-2_17
4. Ciardo, G., Lüttgen, G., Siminiceanu, R.: Saturation: an efficient iteration strategy for symbolic state—space generation. In: Margaria, T., Yi, W. (eds.) TACAS 2001. LNCS, vol. 2031, pp. 328–342. Springer, Heidelberg (2001). https://doi.org/10.1007/3-540-45319-9_23
5. Ciardo, G., Marmorstein, R., Siminiceanu, R.: Saturation unbound. In: Garavel, H., Hatcliff, J. (eds.) TACAS 2003. LNCS, vol. 2619, pp. 379–393. Springer, Heidelberg (2003). https://doi.org/10.1007/3-540-36577-X_27
6. Ciardo, G., Miner, A.S.: SMART: Stochastic Model checking Analyzer for Reliability and Timing, User Manual. http://smart.cs.iastate.edu
7. Ciardo, G., Yu, A.J.: Saturation-based symbolic reachability analysis using conjunctive and disjunctive partitioning. In: Borrione, D., Paul, W. (eds.) CHARME 2005. LNCS, vol. 3725, pp. 146–161. Springer, Heidelberg (2005). https://doi.org/10.1007/11560548_13
8. Cimatti, A., Clarke, E., Giunchiglia, F., Roveri, M.: NuSMV: a new symbolic model verifier. In: Halbwachs, N., Peled, D. (eds.) CAV 1999. LNCS, vol. 1633, pp. 495–499. Springer, Heidelberg (1999). https://doi.org/10.1007/3-540-48683-6_44
9. Couvreur, J.-M., Encrenaz, E., Paviot-Adet, E., Poitrenaud, D., Wacrenier, P.-A.: Data decision diagrams for Petri Net analysis. In: Esparza, J., Lakos, C. (eds.) ICATPN 2002. LNCS, vol. 2360, pp. 101–120. Springer, Heidelberg (2002). https://doi.org/10.1007/3-540-48068-4_8
10. Couvreur, J.-M., Thierry-Mieg, Y.: Hierarchical decision diagrams to exploit model structure. In: Wang, F. (ed.) FORTE 2005. LNCS, vol. 3731, pp. 443–457. Springer, Heidelberg (2005). https://doi.org/10.1007/11562436_32
11. Hamez, A., Thierry-Mieg, Y., Kordon, F.: Hierarchical set decision diagrams and automatic saturation. In: van Hee, K.M., Valk, R. (eds.) PETRI NETS 2008. LNCS, vol. 5062, pp. 211–230. Springer, Heidelberg (2008). https://doi.org/10.1007/978-3-540-68746-7_16
12. Itai, A., Rodeh, M.: Symmetry breaking in distributed networks. In: 22th Annual Symposium on Foundations of Computer Science, pp. 150–158. IEEE Computer Society Press, October 1981
13. Kam, T., Villa, T., Brayton, R.K., Sangiovanni-Vincentelli, A.: Multi-valued decision diagrams: theory and applications. Mult. Valued Log. 4(1–2), 9–62 (1998)
14. Marussy, K., Molnár, V., Vörös, A., Majzik, I.: Getting the priorities right: saturation for prioritised Petri Nets. In: van der Aalst, W., Best, E. (eds.) PETRI NETS 2017. LNCS, vol. 10258, pp. 223–242. Springer, Cham (2017). https://doi.org/10.1007/978-3-319-57861-3_14
15. Minato, S.-i.: Zero-suppressed BDDs and their applications. Softw. Tools Technol. Transf. **3**, 156–170 (2001)

16. Miner, A.S.: Saturation for a general class of models. IEEE Trans. Softw. Eng. **32**(8), 559–570 (2006)
17. Molnár, V., Majzik, I.: Saturation enhanced with conditional locality: application to Petri Nets. In: Donatelli, S., Haar, S. (eds.) PETRI NETS 2019. LNCS, vol. 11522, pp. 342–361. Springer, Cham (2019). https://doi.org/10.1007/978-3-030-21571-2_19
18. Smith, B., Ciardo, G.: SOUPS: a variable ordering metric for the saturation algorithm. In: 2018 18th International Conference on Application of Concurrency to System Design (ACSD), pp. 1–10 (2018)
19. Somenzi, F.: Binary decision diagrams. In: Calculational System Design. NATO Science Series F: Computer and Systems Sciences, vol. 173, pp. 303–366. IOS Press (1999)
20. Wan, M., Ciardo, G.: Symbolic state-space generation of asynchronous systems using extensible decision diagrams. In: Nielsen, M., Kučera, A., Miltersen, P.B., Palamidessi, C., Tůma, P., Valencia, F. (eds.) SOFSEM 2009. LNCS, vol. 5404, pp. 582–594. Springer, Heidelberg (2009). https://doi.org/10.1007/978-3-540-95891-8_52

Case Study: Reachability and Scalability in a Unified Combat-Command-and-Control Model

Sergiy Bogomolov[1,2], Marcelo Forets[3], and Kostiantyn Potomkin[1,2(✉)]

[1] Newcastle University, Newcastle upon Tyne, UK
k.potomkin2@newcastle.ac.uk
[2] Australian National University, Canberra, Australia
[3] Universidad de la República, Montevideo, Uruguay

Abstract. Reachability analysis computes an envelope encompassing the reachable states of a hybrid automaton within a given time horizon. It is known to be a computationally intensive task. In this case study paper, we consider the application of reachability analysis on a mathematical model unifying two key warfighting functions: Combat, and Command-and-Control (C2). Reachability here has a meaning of whether, given a range of initial combat forces and a C2 network and various uncertainties, one side can survive combat with intact forces while the adversary is diminished to zero. These are questions which arise in military Operations Research (OR). This paper is the first to utilize the notions of a hybrid automaton and reachability analysis in the area of OR. We explore the applicability and scalability of Taylor-model based reachability techniques in this domain. Our experiments demonstrate the potential of reachability analysis in the context of OR.

Keywords: Hybrid automata · Reachability analysis · Operations research · Combat · Command and control

1 Introduction

Verification techniques aim at providing mathematically sound guarantees on the correctness of software-intensive [22] or hardware [27] systems. A hybrid automaton [6] is an expressive mathematical model which embeds both discrete and continuous dynamics and thus can be used to model a variety of systems. Reachability analysis [13,15,25] considers properties expressed in terms of reachability of given unsafe system states, distilling them to the computation of the envelope encompassing the system states reachable within a given time horizon. Although some considerable progress in developing and evaluating tools for reachability analysis of hybrid automata with nonlinear dynamics has been achieved [31], there is a need to evaluate these tools on further challenging cases to increase their utility in diverse real-world settings. This case study strives to answer this challenge. Specifically, we consider a model from

© Springer Nature Switzerland AG 2020
S. Schmitz and I. Potapov (Eds.): RP 2020, LNCS 12448, pp. 52–66, 2020.
https://doi.org/10.1007/978-3-030-61739-4_4

Operations Research (OR) and investigate applicability of reachability analysis methods in this domain. We thereby seek to foster the cross-fertilization between Verification and OR.

In more detail, we consider a model based on a unification of two separately well-known mathematical models expressed as differential equations, those of Lanchester and Kuramoto which are invoked separately to model combat between two military forces, and the means by which decisions within them are synchronised. The latter model is fundamentally non-linear creating challenges for reachability tools. In the following, we explore the limits of some hybrid verification tools through the lens of this unified Lanchester-Kuramoto model.

Dating to 1916 [38], the Lanchester model is a representation of combat between two homogeneous forces. While a drastic simplification in many respects, which has not stopped effort at testing it with historical data [50] the model particularly lacks any representation of how decisions are made in the military force. This is where the second model comes in: the Kuramoto model [23] represents agents connected (in its most general form) in some network structure that undertake a limit cycle process while interacting with each other in order to locally synchronise phases [2]. A coupling constant varies the strength of that interaction such that, at high enough strength, spontaneous synchronisation can manifest across the entire network. This model can serve to represent what the military call Command and Control (C2), namely how decision-making should be distributed through a force in order to achieve coherent collective action. In unifying these models, successful synchronisation of decision-making leads to enhanced combat fire power.

The key question in any combat model is: under mutual attrition between two forces of given lethality, over the course of an engagement which side will survive with intact forces? Reachability then becomes the question: with uncertainties about the adversary's initial forces, will one side reach a satisfactory outcome in combat? This paper is the first to showcase the potential of reachability analysis in this context by applying this class of techniques to the unified Lanchester-Kuramoto model. We also demonstrate that hybrid automata are instrumental to encode and analyze the impact of multi-phase strategies employed by the forces during the combat. At the same time, our experimental results indicate that the considered models represent a challenging task for up-to-date reachability analysis techniques. Thus, we hope that the proposed models will be useful for the verification community to guide and calibrate its efforts related to the verification of hybrid automata with nonlinear differential equations.

The paper is organized as follows. In Sect. 2 we recall the mathematical preliminaries needed for this work where we describe two models: Kuramoto and Lanchester, which motivate a definition of unified combat-C2 model. We recall the reachability analysis methods used in the experimental section. A hybrid generalization is also considered. In Sect. 3, we start by defining two classes of network structures that are interesting for this study: hierarchical and fully connected, and then develop the theoretical solution of the model's equations under simplifying assumptions. The experimental evaluation is discussed in Sect. 4. We present conclusions and perspectives for future work in Sect. 5.

2   Preliminaries

2.1   Basic Models

Kuramoto Model. This is a mathematical model representing the phenomenon of synchronisation between phase oscillators coupled on a network. Note that the original model, proposed by Yoshiki Kuramoto in works [36,37], considered N all-to-all coupled oscillators and considered the limit of $N \to \infty$. The state of each oscillator is defined by a variable θ_i. Each oscillator also has its own native frequency ω_i. The model is described by the following ordinary differential equation:

$$\dot{\theta}_i(t) = \omega_i - \sigma \sum_{j=1}^{N} K_{i,j} \sin(\theta_j(t) - \theta_i(t)), \quad i = 1...N \tag{1}$$

where N is the total number of oscillators, σ is the coupling coefficient and $K_{i,j}$ is the *adjacency matrix* of ones and zeroes according to whether nodes i and j are connected or not. Both σ and $K_{i,j}$ control the tightness of organizational coupling, the one through intensity of interaction and the other through the connectivity of that interaction.

The degree of synchronisation is typically measured through a quantity known as the 'order parameter'

$$r(t) = \frac{1}{N} \left| \sum_{j=1}^{N} e^{i\theta_j(t)} \right|, \tag{2}$$

which is bounded $0 \leq r(t) \leq 1$. For low coupling σ, $r(t)$ behaves chaotically with fluctuations of order $1/\sqrt{N}$ indicating the lack of coherence between phases. At large coupling, $r(t)$ reaches a plateau at the value one indicating global synchronisation. The key factors contributing to that degree of synchronisation are the frequencies, the coupling and the connectivity in the adjacency matrix. Note that this model is generally insensitive to initial conditions $\theta_i(0)$: for any set of initial values the transient may vary but not the steady-state (which may be dynamical) behaviour.

This mathematical model has been proposed as useful for a description of C2, namely of a distributed system of decision-makers, such as the staff in a headquarters [35]. These 'C2 agents' can be seen as undertaking a continuous version of the Observe-Orient-Decide-Act (OODA) loop. Originally proposed by US Air Force Colonel John Boyd to identify the key to success in air-to-air combat – agents gain decision advantage to the degree that they succeed in 'getting inside' the OODA loop of the adversary – the model has been generalised to cover the whole spectrum of decision making, from tactical to strategic in both military and civilian endeavours [28,45].

The native frequency of the Kuramoto model thus represents the intrinsic speed with which an agent undertakes their OODA loop. The mathematical model in its OR application [35] then represents distributed OODA agents seeking to locally synchronise decision-making by working within a C2, or organisational, structure and with a given intensity of interaction.

Lanchester Model. Proposed by Frederick Lanchester [38], this is a set of differential equations originally devised to describe the waxing and waning of resources between two opposing military forces. The model is given through the equations

$$\dot{B}(t) = \gamma_b B(t) - \alpha_\rho R(t),$$
$$\dot{R}(t) = \gamma_\rho R(t) - \alpha_b B(t), \tag{3}$$

where one refers to the adversaries as the Blue and Red forces, respectively, and γ and α are resupply and attrition, or lethality, rates respectively, usually taken as constants. In this form of the model, directed fire between the two adversaries is represented, whereas indirect fire (where a side can suffer collateral damage) is modelled by having the attrition terms multiplying both B and R. We consider here the case of direct fire. Subject to various modifications, the Lanchester model is also a mainstay in the OR literature [43].

In some simple cases a conserved quantity may be derived. For example, in the absence of resupply $\gamma_b = \gamma_\rho = 0$ multiplying the top equation by $B(t)/\alpha_\rho$ and the bottom equation by $R(t)/\alpha_b$ and subtracting the two equations from each other yields

$$\frac{d}{dt}(B(t)^2 - R(t)^2) = 0. \tag{4}$$

Thus the difference of squares of the force totals is conserved allowing prediction of the outcome of the battle purely based on initial deployed forces. This is in contrast to the Kuramoto model where initial conditions do not influence the behaviour at large times. In this linear deterministic setting then the reachability of a winning outcome for (say) the Blue force can be answered based on initial conditions.

2.2   The Unified Combat-C2 Model

Continuous Model. The key to unifying C2 into a model of combat lies in the observation that C2 is a 'force multiplier': when C2 is effective the force is able to apply combat power to maximum effect; when C2 is ineffective combat power is undermined, or attenuated. Similarly, in terms of resupply: with effective C2, logistics is effective in enabling flows of resources into combat; with ineffective C2, logistics is undermined and resupply is inadequate.

Based on these principles then, a unified Lanchester-Kuramoto model can be given as [3,34]:

$$\dot{B}(t) = \gamma_B r_B(t)B(t) - \alpha_R r_R(t)R(t)$$
$$\dot{R}(t) = \gamma_R r_R(t)R(t) - \alpha_B r_B(t)B(t) \tag{5}$$

Here $r_B(t)$ and $r_R(t)$ are now order parameters for the C2 systems of the two separate forces. However, since these are both quantities varying between 0 and 1, in view of the additional non-linearity in the absolute value we may use the *squares* of these quantities in the model:

$$r_B^2(t) = \sigma \left| \sum_{j=1}^{N} e^{i\beta_j} \right|^2 = \frac{1}{N^2} \sum_{j,k=1}^{N} \cos(\beta_j - \beta_k)$$

$$r_R^2(t) = \sigma \left| \sum_{j=1}^{N} e^{i\rho_j} \right|^2 = \frac{1}{N^2} \sum_{j,k=1}^{N} \cos(\rho_j - \rho_k). \tag{6}$$

The other part of the dynamics are the separate internal processes of the headquarters of Blue and Red forces, given by separate Kuramoto models

$$\dot{\beta}_i(t) = \omega_i - \sigma_B \sum_{j=1}^{N} B_{i,j} \sin(\beta_j(t) - \beta_i(t)), \quad i = 1...N$$

$$\dot{\rho}_i(t) = \nu_i - \sigma_R \sum_{j=1}^{M} R_{i,j} \sin(\rho_j(t) - \rho_i(t)), \quad i = 1...M. \tag{7}$$

Here N, M represent the number of staff in the headquarters, and we distinguish frequencies and couplings for the two organisations. In this form, though C2 dynamics can enhance or attenuate combat effectiveness there is no feedback of combat losses on the effectiveness of the C2. This is easily achieved by inserting attenuation factors (for example, based on ratios of current to initial forces) in Eqs. (7). In fact such a model is only one step in a considerably generalised model that represents heterogeneity across multiple levels and has been proposed in [3]. We mention the existence of the larger model out, firstly, to emphasise that the present study with a simpler model is a first step in a larger program of applying reachability to military Operations Research models. Secondly, we show this to indicate that despite the apparent simplicity of equations to follow, it would be misleading to develop specific techniques for applying reachability to these simple models that *cannot be extended to the generalised model.*

Returning then to the simpler global Lanchester model with networked C2, for simplicity we shall assume no resupply for the combat forces, namely $\gamma_B = \gamma_R = 0$. The simplified unified combat-C2 model can be described by the following dynamical system:

$$\dot{B}(t) = -\alpha_R r_R^2(t)R(t)$$
$$\dot{R}(t) = -\alpha_B r_B^2(t)B(t). \tag{8}$$

With Eqs. (6) we see that the extra factors in the modified Lanchester equations, Eqs. (8) are purely trigonometric functions. Though the initial force strengths, $B(0), R(0)$ will generally be independent of the size of the C2 systems, N, M, for simplicity of number of inputs in this paper we take them as equal $B(0) = R(0) = N = M$. This is to narrow the scope of the study to *one independent variable* to study scalability, namely the basic network size N. We also choose *initial uncertainties* in these initial values of 0.05, thus $N \leq B(0) \leq N + 0.05$, with a similar assumption for $R(0)$.

Due to the dynamics in the coefficients of this system, the previous conservation considerations no longer hold. Thus, reachability is a considerably more involved process here than only looking at force totals at the initial time, leading to the considerations we pursue in this paper. This model, for all its simplicity, retains one level of network structure – in the Kuramoto dynamics for C2. Thus, when we refer to *scalability* of reachability methods in this paper we mean their performance for increasing N. Consequently, accounting for both Blue and Red, we will be concerned with a $2(N+1)$ dimensional system describing the variables $\beta_i(t), \rho_i(t), B(t), R(t)$.

Hybrid Setting. As mentioned above, one of the features of applying hybrid automata to the combat model is possibility to explore instantaneous changes in some of the parameters at the occurrence of certain threshold conditions, in the course of the engagement. For example, one side may begin with a certain coupling strength and weapons lethality but find itself degraded to a certain point where it needs to make a change, if it is within its capabilities to do so. For example, its headquarters staff may need to interact more intensively with each other. Or it may need to deploy new weapons systems that may have been unavailable at the initial engagement. These can be reflected in step-wise changes in the parameters. Thus the hybrid model may have *two modes*, one mode for each parameter setting.

2.3   Non-linear Reachability

Traditional reachability analysis techniques [14,19,24,47] are applied to systems described by linear or nonlinear Ordinary Differential Equations (ODEs) $\dot{x} = f(x, t)$ along with an initial condition $x \in X_0$ as well as an "unsafe set" U, the purpose being to prove that there is no state reachable contained in U over a bounded (or unbounded) time horizon, for any initial condition in X_0. For nonlinear ODEs, competing approaches are known, such as Taylor-models [12,18], zonotopes [4], polynomial zonotopes [5], interval-based methods [41,44], Hamilton-Jacobi [8] and hybridization-based methods [7]. Several implementations exist, see e.g. [32]. A related problem is the computation of the backwards reachable set [42], for which Taylor model methods have been extended as well [49].

Taylor Models. A Taylor model (TM) of an $(n+1)$-continuously differentiable function $f(x)$ over a given open set $D \in \mathbb{R}^d$ called the domain, is the tuple (p_n, Δ), where p_n is the n-th order polynomial approximation of $f(x)$ and Δ is an interval such that $f(x) \in p_n(x) + \Delta$ for all $x \in D$. This representation was introduced by Berz and Makino [40] and extended by M. Joldes [33]. In particular, TMs have been applied to perform reachability analysis and in this paper we carry our experiments using functionality from the Julia ecosystem [9–11,46], available through a simple interface with the JuliaReach tool [1,14]. Moreover, we ran the case study in another TM-based implementation available in the tool Flow* [19].

3   Selection of the Model Parameters

Network Structures. The Kuramoto model has been applied to a large range of network structures, both regular (complete [16,37], tree [21] and ring [48] graphs) and irregular (small world [29] and scale free [30], often comparing to uniform random or Erdos-Renyi [26]). As said, our aim in this paper is to examine scalability of reachability methods as a function of N. To that end, it suffices that we consider two extreme caricature network structures. Thus we consider: (i) hierarchical organisation for Blue, which for our purposes may be simplified to a path graph (see Leavitt's typology [39]); (ii) flat organisation for Red, namely a fully connected graph. To start small and build up in scale, we shall consider the cases $N = 3, 5$ and $N = 10$.

Random Variables. Initial values for the β_i and ρ_i are chosen typically randomly between 0 and π. Here we choose them such that the i-th elements of Blue and Red have the same values. Similarly, the native frequencies are chosen from some random distribution. In this paper we choose an instance of frequencies uniformly distributed between $(0, 2)$. We show the range of these values for a fixed instance in Fig. 1. We also show here the initial conditions used up to $N = 10$ however, as explained in Sect. 2.1, these play a minimal role in the long term dynamics of the model.

Coupling Strength. Intuitively, hierarchies perform poorly against well-connected structures as they get larger because of the large distances between the apex and lowest levels nodes that are created in the former type of graph. The Kuramoto model quantifies this effect succinctly. Consider the time average of the order parameter

$$\langle r \rangle_T = \frac{1}{(T - T_c)} \int_0^T r(t)dt, \tag{9}$$

where we eliminate a transient for $t < T_c$. This permits computation of $\langle r \rangle_T$ as a function of σ for the two different networks.

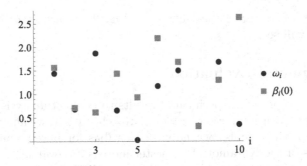

Fig. 1. Random choices of frequencies and initial conditions up to $N = 10$ used in this paper.

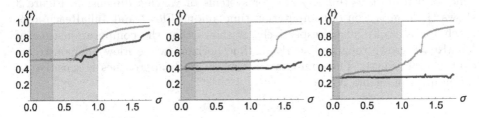

Fig. 2. Time-averaged order parameter for a Blue hierarchy and Red complete graph as a function of coupling, for $N = 3, 5, 10$ from left to right. The shaded regions indicate the values of $\sigma = 1/N$ (light) and $\sigma = 1/N^2$ (dark). (Color figure online)

This is shown in Fig. 2 for the cases $N = 3, 5, 10$ from left to right. To show some indicative values we have shaded the plots to identify the values of $\sigma = 1/N$ (light) and $\sigma = 1/N^2$ (dark). Note our statement earlier that even for pure incoherence $r \approx 1/\sqrt{N}$. Thus for $N = 3$, $r \approx 0.6$, for $N = 5, r \approx 0.4$, $N = 10, r \approx 0.1$, a trend that can be observed in Fig. 2. Hence for small graphs there is already a degree of 'accidental' synchronisation even in the absence of coupling.

But comparing Blue and Red across the plots in Fig. 2, we see that for $N = 3$, at $\sigma = 1/N^2$ Blue and Red are indistinguishable in degree of synchronisation, while at $\sigma = 1/N$ Red shows slightly better synchronisation. In fact, with $\langle r \rangle \approx 0.7$ here this gives Red in fact quite a degree of coherence. This is a consequence of its better connectivity, even for three nodes. For $N = 5$ at both $\sigma = 1/N^2$ and $1/N$ neither side gains in synchronisation, with Red holding an advantage. Finally, for $N = 10$, at the lower value of coupling Red and Blue synchronisable poorly and are indistinguishable, with Blue never really recovering at higher coupling; Red, however, significantly improves at $\sigma = 1/N$. The general trend of the curve of $\langle r \rangle$ for Blue remaining flat is a consequence of the fact that as a path, with N becoming larger and larger, synchronisation becomes increasingly difficult.

In the following then, we examine the unified combat model for coupling $\sigma = 1/N$. Later, in the hybrid setting we will consider switching to coupling

$1/N^2$. But the behaviour of these curves in Fig. 2 enables us to interpret the behaviours we will see.

4   Experimental Evaluation

In this section, we present experimental results to conduct analysis on a Taylor model based approach using Flow* and JuliaReach. Firstly, we analyze the computation time of these tools. Secondly, reach tubes for purely continuous and hybrid versions of the Kuramoto-Lanchester models are evaluated. The models used in the experiments are publicly available on GitHub[1].

Continuous C2 Model. We present the experimental results of computation time with respect to time horizon for systems of varying dimension. Figure 3 shows the comparison of computation time among Flow* and JuliaReach approach for $N = 3$ and 5, respectively. It shows that the computation time of JuliaReach grows roughly linearly in time horizon, while Flow* demonstrates exponential growth. The computation time of both approaches is sensitive to the number of agents N.

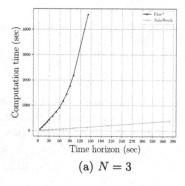
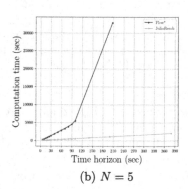

(a) $N = 3$ (b) $N = 5$

Fig. 3. Computation time comparison among Flow* and JuliaReach for $N = 3$ and $N = 5$.

Figure 4 shows the reach tube we construct with JuliaReach for $N = 3$ before either reaching time bound or one of the forces win. σ is equal to $1/N$ and $\alpha = 0.01$ in the experiments. Computational time is 600s and 5800 s for $N = 3$ and $N = 5$ (not shown) respectively. In addition, Figure 4 contains the reach tube for $N = 10$. To compute the reachable set for the system with 10 agents we change the attrition rate α to 0.05, so the flowpipe can be constructed faster. However, it still takes 3.8 h to compute the flowpipe for the 22-dimensional combat model.

In Fig. 4 we also draw a diagonal line representing the trajectory of *equally matched forces*. The reach tube in all cases lies above the diagonal such that

---

[1] https://github.com/kpotomkin/ReachabilityBenchmarks

for all trajectories within the tube Blue reaches zero while Red remains non-zero. Indeed we see that with increasing N the deviation from the diagonal line increases. Here we see for the first time that the reachable set, across values of N, *excludes a Blue victory.* The increasing deviation is a consequence of the increasing superiority of the flat network structure over the hierarchy as N gets larger. This is consistent with Fig. 2 where for coupling $\sigma = 1/N$ Red gains more of an advantage over Blue from its greater superiority in synchronisation – even though for both the degree of synchronisation is generally lower. From Fig. 2, for $N = 10$ both sides generally have poorer synchronisation, however the margin in favour of Red is larger for $N = 10$ than it is for smaller N. These are consequences of the superior network structure of Red. However, we also observe that the fluctuations with $N = 10$ are larger than those for smaller N. This is also consistent with Fig. 2 where, for the right most plot the overall degree of synchronisation $\langle r \rangle$ is lower for larger N; larger graphs of the same type require more coupling to obtain the corresponding degree of synchronisation for the smaller graph. This poorer overall synchronisation is then reflected in greater fluctuations in the combat dynamics shown for $N = 10$.

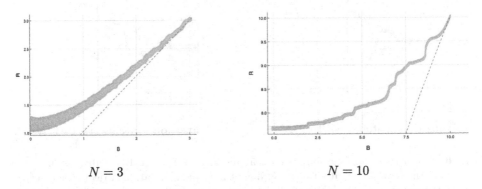

Fig. 4. Reach tubes (orange) along with simulation trajectories (green) up to either reaching time bound or one of the forces win, i.e. the other reaches zero. Blue dash line corresponds to the function $y(x) = x$. (Color figure online)

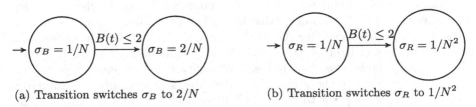

(a) Transition switches σ_B to $2/N$ (b) Transition switches σ_R to $1/N^2$

Fig. 5. Blue forces affect coupling between agents when below a critical threshold. (Color figure online)

Hybrid Combat Model. We alluded at the outset that the hybrid automata approach permits examination of multiple modes. Evidently in the previous results, the Blue force suffers higher attrition as a consequence of its poorer network structure. Arguably, that force has the capacity to increase coupling within its structure upon realising it is degrading faster than the adversary. For example, we choose the mode switch to be triggered by the value of B reaching the threshold value of 2, as illustrated in Fig. 5. Flow* does not manage to handle discrete transitions in the proposed models. Therefore, the results only for JuliaReach are presented below.

In Fig. 6 we show the result of the hybrid model for $N = 3$. We can observe, that instead of losing the engagement (red set), doubling the tightness of organizational coupling (the value of σ) shifts the flow of the battle, so that the Blue force eventually wins the engagement (orange). Again, the reach tube for the Blue force after its switches mode is entirely in the victory region, whereby Red is zero for all positive values of Blue.

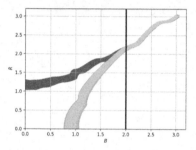   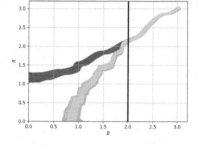

Transition switches σ_B to $2/N$ Transition switches σ_R to $1/N^2$

Fig. 6. Reach tubes for the hybrid combat models for $N = 3$. Red tubes assume no change in the strategy of either force. Orange tubes and simulation trajectories (green) reflect the changes of the tightness of organizational coupling at the position indicated by the vertical line, such as increasing in Blue forces (left) and decreasing in Red forces (right). (Color figure online)

Alternately, Blue may undertake some action to disrupt the connection strength between Red force elements, such that, for example, σ_R is reduced by a factor of $1/N$, as illustrated in Fig. 5. Examining the reach tubes, again for $N = 3$ shown in Fig. 6, we observe that instead of losing Blue guarantees victory across all trajectories evolving out of the initial set. Again, we can understand these behaviours in light of Fig. 2. For $N = 3$, by doubling its coupling Blue gains an appreciable advantage in terms of ability to synchronise: it goes from $\langle r \rangle \approx 0.6$ up to close to the value of one, with Red still at nearly 0.7. Similarly, weakening Red's coupling to $\sigma = 1/N^2$ drops Red's average synchronisation from 0.7 down to approximately 0.5. Either way, this demonstrates in principle how hybrid automata may address questions around change in properties of a system during the course of a dynamical process.

5   Conclusion

We have explored the feasibility of applying some state-of-the-art reachability tools on a mathematical model unifying the dynamics of combat and C2. Operations Research is a domain where the advantages of using reachability analysis are yet to be discovered, and our contribution is a first step towards that direction. The unified combat model considered appears to be a challenging case study model, and although current verification techniques demonstrate capabilities to verify the systems up to the necessary time horizon of the model, there are still relevant configurations where reachability tools fail to verify the system.

In particular, we investigated the method based on Taylor models in Flow* and JuliaReach for cases where the C2 takes place on networks of size $N = 3, 5$ and 10. Given a certain uncertainty on initial conditions for Blue and Red combat forces we were able to derive reach-tubes for the subsequent time-evolution, bounding the behaviours of both forces and determining whether one side or the other is guaranteed success in the combat outcome. In conclusion, the approach with adaptive time step (JuliaReach) has shown the best scalability. We emphasise that here we examined reachability within the constraint that the reach-tubes should cover all possible behaviors of the system's evolution. However, we anticipate that significant scaling benefits may be achieved by relaxing this condition, thus testing that certain states may be reached while allowing for a degree of approximation.

Additional future directions are various. For example, employing techniques to further exploit the model structure, e.g. decomposition [20]. There are also techniques available to cope with nonlinearities more accurately, such as *clustering* to reduce the approximation error caused by successive convexification and overapproximation, or more sophisticated methods to reduce the wrapping effect [17]. The hybrid version of the model here posed challenges for the verification engine, in the main, because of the intersection and clustering steps. Therefore, this case study is an appealing model to test new techniques and highlight essential research directions, such as refined flowpipe-guard intersection methods.

Acknowlegements. The authors would like to thank Alexander C. Kalloniatis from Joint and Operations Analysis Division, Defence Science and Technology Group for many productive discussions.

This research was collaboration between the Commonwealth of Australia represented by the Defence Science and Technology Group and Australian National University, where this work was initiated, through a Defence Science Partnerships agreement. The research was conducted under the auspices of the Modelling Complex Warfighting initiative and was supported in part by the Air Force Office of Scientific Research under award number FA2386-17-1-4065. Any opinions, findings, and conclusions or recommendations expressed in this material are those of the authors and do not necessarily reflect the views of the United States Air Force.

References

1. JuliaReach. https://github.com/JuliaReach (2017)
2. Acebrón, J.A., Bonilla, L.L., Vicente, C.J.P., Ritort, F., Spigler, R.: The Kuramoto model: a simple paradigm for synchronization phenomena. Rev. Modern Phys. **77**(1), 137 (2005)
3. Ahern, R., Zuparic, M., Kalloniatis, A., Hoek, K.: Unifying warfighting functions in mathematical modelling: combat, Manoeuvre and C2. Submitted to Journal of the Operational research Society (JORS)
4. Althoff, M.: Reachability analysis and its application to the safety assessment of autonomous cars. Ph.D. thesis, Technische Universität München (2010)
5. Althoff, M.: Reachability analysis of nonlinear systems using conservative polynomialization and non-convex sets. In: Proceedings of the 16th International Conference on Hybrid Systems: Computation and Control, pp. 173–182. ACM (2013)
6. Alur, R., Courcoubetis, C., Henzinger, T.A., Ho, P.-H.: Hybrid automata: an algorithmic approach to the specification and verification of hybrid systems. In: Grossman, R.L., Nerode, A., Ravn, A.P., Rischel, H. (eds.) HS 1991-1992. LNCS, vol. 736, pp. 209–229. Springer, Heidelberg (1993). https://doi.org/10.1007/3-540-57318-6_30
7. Bak, S., Bogomolov, S., Henzinger, T.A., Johnson, T.T., Prakash, P.: Scalable static hybridization methods for analysis of nonlinear systems. In: 19th International Conference on Hybrid Systems: Computation and Control (HSCC 2016), pp. 155–164. ACM
8. Bansal, S., Chen, M., Herbert, S., Tomlin, C.J.: Hamilton-Jacobi reachability: a brief overview and recent advances. In: IEEE 56th Annual Conference on Decision and Control (CDC), pp. 2242–2253. IEEE (2017)
9. Benet, L., Sanders, D.: TaylorSeries.jl: Taylor expansions in one and several variables in Julia. J. Open Source Softw. **4**, 1043 (2019)
10. Benet, L., Sanders, D.P.: JuliaDiff/TaylorSeries.jl, March 2019. https://doi.org/10.5281/zenodo.2601942
11. Benet, L., Sanders, D.P.: JuliaIntervals/TaylorModels.jl, March 2019. https://doi.org/10.5281/zenodo.2613103
12. Berz, M., Makino, K.: Verified integration of ODEs and flows using differential algebraic methods on high-order Taylor models. Reliable Comput. **4**(4), 361–369 (1998). https://doi.org/10.1023/A:1024467732637
13. Bogomolov, S., et al.: Guided search for hybrid systems based on coarse-grained space abstractions. Int. J. Softw. Tools Tech. Trans. **18**(4), 449–467 (2015). https://doi.org/10.1007/s10009-015-0393-y
14. Bogomolov, S., Forets, M., Frehse, G., Potomkin, K., Schilling, C.: JuliaReach: a toolbox for set-based reachability. In: 22nd ACM International Conference on Hybrid Systems: Computation and Control (HSCC 2019), pp. 39–44. ACM (2019)
15. Bogomolov, S., Mitrohin, C., Podelski, A.: Composing reachability analyses of hybrid systems for safety and stability. In: Bouajjani, A., Chin, W.-N. (eds.) ATVA 2010. LNCS, vol. 6252, pp. 67–81. Springer, Heidelberg (2010). https://doi.org/10.1007/978-3-642-15643-4_7
16. Bronski, J., deVille, L., Park, M.J.: Fully synchronous solutions and the synchronization phase transition for the finite-N Kuramoto model. Chaos **22**(3), 033133 (2012)
17. Bünger, F.: Shrink wrapping for Taylor models revisited. Numer. Algorithms **78**(4), 1001–1017 (2017). https://doi.org/10.1007/s11075-017-0410-1

18. Chen, X., Abraham, E., Sankaranarayanan, S.: Taylor model flowpipe construction for non-linear hybrid systems. In: IEEE 33rd Real-Time Systems Symposium, pp. 183–192. IEEE (2012)
19. Chen, X., Ábrahám, E., Sankaranarayanan, S.: Flow*: an analyzer for non-linear hybrid systems. In: Sharygina, N., Veith, H. (eds.) CAV 2013. LNCS, vol. 8044, pp. 258–263. Springer, Heidelberg (2013). https://doi.org/10.1007/978-3-642-39799-8_18
20. Chen, X., Sankaranarayanan, S.: Decomposed reachability analysis for nonlinear systems. In: IEEE Real-Time Systems Symposium (RTSS), pp. 13–24. IEEE (2016)
21. Dekker, A., Taylor, R.: Synchronization properties of trees in the Kuramoto model. SIAM J. Appl. Dyn. Sys. **12**(2), 596–617 (2013)
22. D'silva, V., Kroening, D., Weissenbacher, G.: A survey of automated techniques for formal software verification. IEEE Trans. Comput. Aided Des. Integr. Circuits Syst. **27**(7), 1165–1178 (2008)
23. da Fonseca, J., Abud, C.: The Kuramoto model revisited. J. Stat. Mech: Theory Exp. **2018**(10), 103204 (2018)
24. Frehse, G., et al.: SpaceEx: scalable verification of hybrid systems. In: Gopalakrishnan, G., Qadeer, S. (eds.) CAV 2011. LNCS, vol. 6806, pp. 379–395. Springer, Heidelberg (2011). https://doi.org/10.1007/978-3-642-22110-1_30
25. Girard, A., Guernic, C.L.: Efficient reachability analysis for linear systems using support functions. IFAC Proc. Vol. **41**, 8966–8971 (2008)
26. Gomez-Gardenes, J., Moreno, Y., Arenas, A.: Synchronizability determined by coupling strengths and topology on complex networks. Phys. Rev. E **75**, 066106 (2007)
27. Gupta, A.: Formal hardware verification methods: a survey. Form Method Syst. Des. **1**, 151–238 (1992). In: Computer-Aided Verification. pp. 5–92. Springer
28. Hasík, J.: Beyond the briefing: theoretical and practical problems in the works and legacy of John Boyd. Contemp. Secur. Policy **34**(3), 583–599 (2013)
29. Hong, H., Choi, M.Y., Kim, B.J.: Synchronization on small-world networks. Phys. Rev. E **65**(2), 026139 (2002)
30. Ichinomiya, T.: Frequency synchronization in a random oscillator network. Phys. Rev. E **70**(2), 026116 (2004)
31. Immler, F., et al.: ARCH-COMP19 category report: Continuous and hybrid systems with nonlinear dynamics. In: ARCH19. 6th International Workshop on Applied Verification of Continuous and Hybrid Systemsi, part of CPS-IoT Week 2019, Montreal, QC, Canada, pp. 41–61 (2019)
32. Immler, F., et al.: ARCH-COMP19 category report: continuous and hybrid systems with nonlinear dynamics. EPiC Ser. Comput. **61**, 41–61 (2019)
33. Joldes, M.M.: Rigorous polynomial approximations and applications. Ph.D. thesis (2011)
34. Kalloniatis, A., Hoek, K., Zuparic, M.: Network synchronisation and next generation combat models - a dynamical systems approach. In: 86th Military Operations Research Society Symposium (2018)
35. Kalloniatis, A., McLennan-Smith, T., Roberts, D.: Modelling distributed decision-making in command and control using stochastic network synchronisation. Eur. J. Oper. Res. (2020). https://doi.org/10.1016/j.ejor.2019.12.033
36. Kuramoto, Y.: International Symposium on Mathematical Problems in Theoretical Physics. Lecture Notes in Physics, p. 420. Springer, Heidelberg (1975). https://doi.org/10.1007/BFb0013294
37. Kuramoto, Y.: Chemical Oscillations, Waves, and Turbulence. Courier Corporation (2003)

38. Lanchester, F.W.: Aircraft in Warfare: The Dawn of the Fourth Arm. Constable limited (1916)
39. Leavitt, H.J.: Some effects of certain communication patterns on group performance. J. Abnorm. Soc. Psychol. **46**(1), 38–50 (1951)
40. Makino, K., Berz, M.: Taylor models and other validated functional inclusion methods. Int. J. Pure Appl. Math. **6**, 239–316 (2003)
41. Meyer, P.J., Devonport, A., Arcak, M.: Tira: toolbox for interval reachability analysis. In: Proceedings of the 22nd ACM International Conference on Hybrid Systems: Computation and Control, pp. 224–229. ACM (2019)
42. Mitchell, I.M.: Comparing forward and backward reachability as tools for safety analysis. In: Bemporad, A., Bicchi, A., Buttazzo, G. (eds.) HSCC 2007. LNCS, vol. 4416, pp. 428–443. Springer, Heidelberg (2007). https://doi.org/10.1007/978-3-540-71493-4_34
43. Morse, P., Kimball, G.: Methods of Operations Research. Massachusetts Institute of Technology (1951)
44. Nedialkov, N.S.: Interval tools for ODEs and DAEs. In: 12th GAMM-IMACS International Symposium on Scientific Computing, Computer Arithmetic and Validated Numerics (SCAN 2006), p. 4. IEEE (2006)
45. Osinga, F.: "Getting" a discourse on winning and losing: a primer on Boyd's "theory of intellectual evolution". Contemp. Secur. Policy **34**(3), 603–624 (2013)
46. Pérez-Hernández, J.A., Benet, L.: Perezhz/taylorintegration.jl, February 2019. https://doi.org/10.5281/zenodo.2562353
47. Ray, R., Gurung, A., Das, B., Bartocci, E., Bogomolov, S., Grosu, R.: XSpeed: accelerating reachability analysis on multi-core processors. In: Piterman, N. (ed.) HVC 2015. LNCS, vol. 9434, pp. 3–18. Springer, Cham (2015). https://doi.org/10.1007/978-3-319-26287-1_1
48. Rogge, J.A., Aeyals, D.: Stability of phase locking in a ring of unidirectionally coupled oscillators. SIAM J. Appl. Dyn. Syst. **37**, 11135–11148 (2004)
49. Rwth, X.C., Sankaranarayanan, S., Ábrahám, E.: Under-approximate flowpipes for non-linear continuous systems. In: Formal Methods in Computer-Aided Design (FMCAD), pp. 59–66. IEEE (2014)
50. Tam, J.H.: Application of Lanchester combat model in the Ardennes campaign. Nat. Resour. Model. **11**(2), 95–116 (1998)

Qualitative Multi-objective Reachability for Ordered Branching MDPs

Kousha Etessami$^{(\boxtimes)}$ and Emanuel Martinov$^{(\boxtimes)}$

School of Informatics, University of Edinburgh, Edinburgh, UK
kousha@inf.ed.ac.uk, eo.martinov@gmail.com

Abstract. We study qualitative multi-objective reachability problems for Ordered Branching Markov Decision Processes (OBMDPs), or equivalently context-free MDPs, building on prior results for single-target reachability on Branching Markov Decision Processes (BMDPs).

We provide two separate algorithms for "almost-sure" and "limit-sure" multi-target reachability for OBMDPs. Specifically, given an OBMDP, $\mathcal{A}$, given a starting non-terminal, and given a set of *target* non-terminals K of size $k = |K|$, our first algorithm decides whether the supremum probability, of generating a tree that contains every target non-terminal in set K, is 1. Our second algorithm decides whether there is a strategy for the player to almost-surely (with probability 1) generate a tree that contains every target non-terminal in set K. The two separate algorithms are needed: we give examples showing that indeed "almost-sure" $\neq$ "limit-sure" for multi-target reachability in OBMDPs. Both algorithms run in time $2^{O(k)} \cdot |\mathcal{A}|^{O(1)}$, where $|\mathcal{A}|$ is the bit encoding length of $\mathcal{A}$. Hence they run in P-time when k is fixed, and are fixed-parameter tractable with respect to k. Moreover, we show that the qualitative almost-sure (and limit-sure) multi-target reachability decision problem is in general NP-hard, when k is not fixed.

Keywords: Markov decision processes · Branching processes · Stochastic context-free grammars · Multi-objective · Reachability · Almost-sure · Limit-sure

1 Introduction

Ordered Branching Markov Decision Processes (OBMDPs) can be viewed as controlled/probabilistic context-free grammars, but without any terminal symbols, and where moreover the non-terminals are partitioned into two sets: controlled non-terminals and probabilistic non-terminals. Each non-terminal, N, has an associated set of grammar rules of the form $N \rightarrow \gamma$, where γ is a (possibly empty) sequence of non-terminals. Each probabilistic non-terminal is equipped with a given probability distribution on its associated grammar rules. For each controlled non-terminal, M, there is an associated non-empty set of available

---

A full version [10] of this paper is available at arXiv:2008.10591.

actions, A_M, which is in one-to-one correspondence with the grammar rules of M. So, for each action, $a \in A_M$, there is an associated grammar rule $M \xrightarrow{a} \gamma$. Given an OBMDP, given a "start" non-terminal, and given a "strategy" for the controller, these together determine a probabilistic process that generates a (possibly infinite) random ordered tree. The tree is formed via the usual parse tree expansion of grammar rules, proceeding generation by generation, in a top-down manner. Starting with a root node labeled by the "start" non-terminal, the ordered tree is generated based on the controller's (possibly randomized) choice of action at each node of the tree that is labeled by a controlled non-terminal, and based on the probabilistic choice of a grammar rule at nodes that are labeled by a probabilistic non-terminal.

We assume that a general *strategy* for the controller can operate as follows: at each node v of the ordered tree, labeled by a controlled non-terminal, the controller (player) can choose its action (or its probability distribution on actions) at v based on the entire "ancestor history" of v, meaning based on the entire sequence of labeled nodes and actions leading from the root node to v, *as well as* based on the ordered position of each of its ancestors (including v itself) among its siblings in the tree.

Ordered Branching Processes (OBPs) are OBMDPs without any controlled non-terminals. Both OBPs and OBMDPs are very similar to classic multi-type branching processes (BPs), and to Branching MDP (BMDPs), respectively. The only difference is that for OB(MD)Ps the generated tree is *ordered*. In particular, the rules for an OBMDP have an ordered *sequence* of non-terminals on their right hand side, whereas there is no such ordering in BPs or BMDPs: each rule for a given type associates an unordered multi-set of "offsprings" of various types to that given type. Branching processes and stochastic context-free grammars have well-known applications in many fields, including in natural language processing, biology/bioinformatics (e.g., [16], population genetics [15], RNA modeling [5], and cancer tumor growth modelling [1,19]), and physics (e.g., nuclear chain reactions). Generalizing these models to MDPs is natural, and can allow us to study, and to optimize algorithmically, settings where such random processes can partially be controlled.

The single-target reachability objective for OBMDPs amounts to optimizing (maximizing or minimizing) the probability that, starting at a given start (root) non-terminal, the generated tree contains some given target non-terminal. This objective has already been thoroughly studied for BMDPs, as well as for (concurrent) stochastic game generalizations of BMDPs [8,9]. Moreover, it turns out that there is really no difference at all between BMDPs and OBMDPs when it comes to the single-target reachability objective: all the algorithmic results from [8,9] carry over, mutatis mutantis, for OBMDPs, and for their stochastic game generalizations.

A natural generalization of single-target reachability is multi-objective reachability, where the goal is to optimize each of the respective probabilities that the generated tree contains each of several different target non-terminals. (Of course, there may be trade-offs between these different objectives.)

Our main concern in this paper is *qualitative* multi-objective reachability problems, where the aim is to determine whether there is a strategy that guarantees that each of the given set of target non-terminals is almost-surely (respectively, limit-surely) contained in the generated tree, i.e., with probability 1 (respectively, with probability arbitrarily close to 1). In fact, we show that the *almost-sure* and *limit-sure* problems do not coincide. That is, there are OBMDPs for which there is no single strategy that achieves probability exactly 1 for reaching all targets, but where nevertheless, for every $\epsilon > 0$, there is a strategy that guarantees a probability $\geq 1 - \epsilon$, of reaching all targets.

By contrast, for both BMDPs and OBMDPs, for single-target reachability, the *qualitative* almost-sure and limit-sure questions do coincide [8].[1]

We give two separate algorithms for almost-sure and limit-sure multi-objective reachability. For the *almost-sure* problem, we are given an OBMDP, a start non-terminal, and a set of target non-terminals, and we must decide whether there exists a strategy using which the process generates, with probability 1, a tree that contains all the given target non-terminals. If the answer is "yes", the algorithm can also construct a (randomized) witness strategy that achieves this.[2] The algorithm for the *limit-sure* problem decides whether the supremum probability of generating a tree that contains all given target non-terminals is 1. If the answer is "yes", the algorithm can also construct, given any $\epsilon > 0$, a randomized non-static strategy that guarantees probability $\geq 1 - \epsilon$. The limit-sure algorithm is only slightly more involved.

Both algorithms run in time $2^{O(k)} \cdot |\mathcal{A}|^{O(1)}$, where $|\mathcal{A}|$ is the total bit encoding length of the given OBMDP, $\mathcal{A}$, and $k = |K|$ is the size of the given set K of target non-terminals. Hence they run in polynomial time when k is fixed, and are fixed-parameter tractable with respect to k. Moreover, we show that the qualitative almost-sure (and limit-sure) multi-target reachability decision problem is in general NP-hard, when k is not fixed.

We leave open the decidability of arbitrary boolean combinations of qualitative reachability and non-reachability queries over different target non-terminals. (See the full version for an elaboration on such questions, and algorithms for

---

[1] The notion of general "strategy" employed for BMDPs in [8] is somewhat different than what we define in this paper for OBMDPs: it allows the controller to not only base its choice at a tree node on the ancestor chain of that node, but on the entire tree up to that "generation". This is needed for BMDPs because there is no ordering available on "siblings" in the tree generated by a BMDP. However, a careful look shows that the results of [8] imply that, for OBMDPs, for single-target reachability, almost-sure and limit-sure reachability also coincide under the notion of "strategy" we have defined in this paper, where choices are based only on the "ancestor history" (with ordering information) of each node in the *ordered* tree. In particular the key"queen/workers" strategy employed for almost-sure (=limit-sure) reachability in [8] can be mimicked using the ordering with respect to siblings that is available in ancestor histories of OBMDPs.

[2] This strategy is, however, necessarily not "static", meaning it must actually use the ancestor history: the action distribution cannot be defined solely based on which non-terminal is being expanded.

some special cases.) Furthermore, we leave open all (both decision and approximation) *quantitative* multi-objective reachability questions, including when the goal is to approximate the tradeoff *pareto curve* of optimal probabilities for different reachability objectives. These are intriguing questions for future research.

Related Work. As already mentioned, the single-target reachability problem for OBMDPs (and its stochastic game generalization) is equivalent to the same problem for BMDPs, and was studied in detail in [8,9], even in the quantitative sense. The same holds for another fundamental objective, namely *termination/extinction*, i.e., where the objective is to optimize the probability that the generated tree is finite. The extinction objective for BPs and BMDPs, and the closely related model of 1-exit recursive MDPs, was thoroughly studied in [6,7,12–14], including both qualitative and quantitative algorithmic questions.

Algorithms for checking other properties of BPs and BMDPs have also been investigated before, some of which generalize termination and reachability. In particular, model checking of BPs with properties given by a deterministic parity tree automaton was studied in [3], and in [17] for properties represented by a subclass of alternating parity tree automata. More recently, [18] investigated the determinacy and the complexity of decision problems for ordered branching simple (turn-based) stochastic games with respect to properties defined by finite tree automata defining regular languages on infinite trees. They showed that (unlike the case with reachability) already for some basic regular properties these games are not even determined, meaning they do not have a value. Moreover, they show that for what amounts to OBMDPs with a regular tree objective it is undecidable to compare the optimal probability to a threshold value. Their results do not have implications for (neither quantitative nor qualitative) multi-objective reachability.

Multi-objective reachability and model checking (with respect to omega-regular properties) has been studied for finite-state MDPs in [11], both with respect to qualitative and quantitative problems. In particular, it was shown in [11] that for multi-objective reachability in finite-state MDPs, memoryless (but randomized) strategies are sufficient, that both qualitative and quantitative multi-objective reachability queries can be decided in P-time, and the *Pareto curve* for them can be approximated within a desired error $\epsilon > 0$ in P-time in the size of the MDP and $1/\epsilon$.

Due to space limits, most proofs are omitted (see the full version [10]).

2   Definitions and Background

Rather than providing the most general possible definition of OBMDPs, where rules can have an arbitrarily long string of non-terminals on their right hand side (RHS), to simplify matters, we assume OBMDPs are already in a "simple normal form". This is entirely without loss of generality for our purposes: any OBMDP can be converted efficiently to an "equivalent"[3] one in normal form. This is

---

[3] Equivalent w.r.t. all (multi-objective) reachability objectives we consider.

directly analogous to standard normal form results for context-free grammars, and to similar prior results established for BMDPs [8].

Definition 1. *An Ordered Branching Markov Decision Process (OBMDP), $\mathcal{A}$, (in simple normal form (SNF)) is represented by a tuple $\mathcal{A} = (V, \Sigma, \Gamma, R)$, where $V = \{T_1, \ldots, T_n\}$ is a finite set of non-terminals, and Σ is a finite non-empty action alphabet. The set of non-terminals V is partitioned into three possible kinds: "controlled" (M-Form) non-terminals, "linear (probabilistic)" (L-Form) non-terminals, and "quadratic (branching)" (Q-Form) non-terminals. For each controlled non-terminal T_i, $\Gamma^i \subseteq \Sigma$ is a non-empty set of actions available for T_i. R defines, for each non-terminal $T_i \in V$, a set of (probabilistic/controlled) rules $R(T_i)$.*

Specifically, the set of rules $R(T_i)$ associated with non-terminal $T_i \in V$, has the following structure, depending on what form (kind) of non-terminal T_i is:

- L-FORM: *T_i is a "linear" or "probabilistic" non-terminal, the player has no choice of actions, and the associated rules for T_i are given by: $T_i \xrightarrow{p_{i,0}} \varnothing$, $T_i \xrightarrow{p_{i,1}} T_1, \ldots, T_i \xrightarrow{p_{i,n}} T_n$, where for all $0 \leq j \leq n$, $p_{i,j} \geq 0$ denotes the probability of each rule, and $\sum_{j=0}^{n} p_{i,j} = 1$.*
- Q-FORM: *T_i is a "quadratic" (or "branching") non-terminal, with a single associated rule (and no associated actions), of the form $T_i \xrightarrow{1} T_j T_{j'}$.*
- M-FORM: *T_i a "controlled" non-terminal, with a non-empty set of associated actions $\Gamma^i = \{a_1, \ldots, a_{m_i}\} \subseteq \Sigma$, and the associated rules have the form $T_i \xrightarrow{a_1} T_{j_1}, \ldots, T_i \xrightarrow{a_{m_i}} T_{j_{m_i}}$.[4]*

We denote by $|\mathcal{A}|$ the total bit encoding length of the OBMDP, where we assume the given rule probabilities are rational numbers represented as usual (with numerator and denominator in binary). If $|\Gamma^i| = 1$ for all controlled non-terminals $T_i \in V$ (meaning the controller has no choices), then the model is an *Ordered Branching Process (OBP)*.

A *derivation* for an OBMDP, starting at some start non-terminal $T_{start} \in V$, is a (possibly infinite) labeled ordered tree, $X = (B, s)$, defined as follows. The set of nodes $B \subseteq \{l, r, u\}^*$ of the tree, X, is a *prefix-closed* subset of $\{l, r, u\}^*$.[5] So each node in B is a string over $\{l, r, u\}$, and if $w = w'a \in B$, where $a \in \{l, r, u\}$, then $w' \in B$. As usual, when $w \in B$ and $w' = wa \in B$, for some $a \in \{l, r, u\}$, we call w the *parent* of w', and we call w' a *child* of w in the tree. A *leaf* of B is a node $w \in B$ that has no children in B. Let $\mathcal{L}_B \subseteq B$ denote the set of all leaves in B. The *root* node is the empty string ε (note that B is prefix-closed, so $\varepsilon \in B$). The function $s : B \to V \cup \{\varnothing\}$ assigns either a non-terminal or the empty symbol as a label to each node of the tree, and must satisfy the following conditions: Firstly, $s(\varepsilon) = T_{start}$, in other words the root must be labeled by the start non-terminal; Inductively, if for any *non-leaf* node $w \in B \setminus \mathcal{L}_B$ we have $s(w) = T_i$, for some $T_i \in V$, then:

---

[4] We assume, without loss of generality, that for $0 \leq t < t' \leq m_i$, $T_{j_t} \neq T_{j_{t'}}$.
[5] Here 'l', 'r', and 'u', stand for 'left', 'right', and 'unique' child, respectively.

- if T_i is a Q-form (branching) non-terminal, whose associated unique rule is $T_i \xrightarrow{1} T_j \, T_{j'}$, then w must have exactly two children in B, namely $wl \in B$ and $wr \in B$, and moreover we must have $s(wl) = T_j$ and $s(wr) = T_{j'}$.
- if T_i is a L-form (linear/probabilistic) non-terminal, then w must have exactly one child in B, namely wu, and it must be the case that either $s(wu) = T_j$, where there exists some rule $T_i \xrightarrow{p_{i,j}} T_j$ with a positive probability $p_{i,j} > 0$, or else $s(wu) = \varnothing$, where there exists a rule $T_i \xrightarrow{p_{i,0}} \varnothing$, with an empty right hand side, and a positive probability $p_{i,0} > 0$.
- if T_i is a M-form (controlled) non-terminal, then w must have exactly one child in B, namely wu, and it must be the case that $s(wu) = T_{j_t}$, where there exists some rule $T_i \xrightarrow{a_t} T_{j_t}$, associated with some action $a_t \in \Gamma^i$, having non-terminal T_i as its left hand side.

A derivation $X = (B, s)$ is *finite* if the set B is finite. A derivation $X' = (B', s')$ is called a *subderivation* of a derivation $X = (B, s)$, if $B' \subseteq B$ and $s' = s|_{B'}$ (i.e., s' is the function s, restricted to the domain B'). We use $X' \preceq X$ to denote the fact that X' is a subderivation of X.

A *complete* derivation, or a *play*, $X = (B, s)$, is by definition a derivation in which for all leaves $w \in \mathcal{L}_B$, $s(w) = \varnothing$. For a play $X = (B, s)$, and a node $w \in B$, we define the *subplay of X rooted at w*, to be the play $X^w = (B^w, s^w)$, where $B^w = \{w' \in \{l, r, u\}^* \mid ww' \in B\}$ and $s^w : B^w \to V \cup \{\varnothing\}$ is given by, $s^w(w') := s(ww')$ for all $w' \in B^w$.[6] Consider any derivation $X = (B, s)$, and any node $w = w_1 \ldots w_m \in B$, where $w_k \in \{l, r, u\}$ for all $k \in [m]$. We define the *ancestor history* of w to be a sequence $h_w \in V(\{l, r, u\} \times V)^*$, given by $h_w := s(\varepsilon)(w_1, s(w_1))(w_2, s(w_1 w_2))(w_3, s(w_1 w_2 w_3)) \ldots (w_m, s(w_1 w_2 \ldots w_m))$. In other words, the ancestor history h_w of node w specifies the sequence of moves that determine each ancestor of w (starting at ε and including w itself), and also specifies the sequence of non-terminals that label each of ancestor of w.

For an OBMDP, $\mathcal{A}$, a sequence $h \in V(\{l, r, u\} \times V)^*$ is called a *valid* ancestor history if there is some derivation $X = (B', s')$ of $\mathcal{A}$, and node $w \in B'$ such that $h = h_w$. We define the *current non-terminal* of such a valid ancestor history h to be $s'(w)$. In other words, it is the non-terminal that labels the last node of the ancestor history h. Let $\mathrm{current}(h)$ denote the current non-terminal of h. Let $H_\mathcal{A} \subseteq V(\{l, r, u\} \times V)^*$ denote the set of all valid ancestor histories of $\mathcal{A}$. A valid ancestor history $h \in H_\mathcal{A}$ is said to *belong to the controller*, if $\mathrm{current}(h)$ is a M-form (controlled) non-terminal. Let $H_\mathcal{A}^C$ denote the set of all valid ancestor histories of the OBMDP, $\mathcal{A}$, that belong to the controller.

For an OBMDP, $\mathcal{A}$, a *strategy* for the controller is a function, $\sigma : H_\mathcal{A}^C \to \Delta(\Sigma)$ from the set of valid ancestor histories belonging to the controller, to probability distributions on actions, such that moreover for any $h \in H_\mathcal{A}^C$, if $\mathrm{current}(h) = T_i$, then $\sigma(h) \in \Delta(\Gamma^i)$. (In other words, the probability distribution must have

---

[6] To avoid confusion, note that subderivation and subplay have very different meanings. Saying derivation X is a "subderivation" of X', means that in a sense X is a "prefix" of X', as an ordered tree. Saying play X is a subplay of play X', means X is a "suffix" of X', more specifically X is a subtree rooted at a specific node of X'.

support only on the actions available at the current non-terminal.) Note that the strategy can choose different distributions on actions at different occurrences of the same non-terminal in the derivation tree, even when these occurrences happen to be "siblings" in the tree.

Let Ψ be the set of all strategies. We say $\sigma \in \Psi$ is *deterministic* if for all $h \in H_{\mathcal{A}}^C$, $\sigma(h)$ puts probability 1 on a single action. We say $\sigma \in \Psi$ is *static* if for each M-form (controlled) non-terminal T_i, there is some distribution $\delta_i \in \Delta(\Gamma^i)$, such that for any $h \in H_{\mathcal{A}}^C$ with $\mathtt{current}(h) = T_i$, $\sigma(h) = \delta_i$. In other words, a static strategy σ plays exactly the same distribution on actions at every occurrence of each non-terminal T_i, regardless of the ancestor history.

For an OBMDP, $\mathcal{A}$, fixing a start non-terminal T_i, and fixing a strategy σ for the controller, determines a stochastic process that generates a random play, as follows. The process generates a sequence of finite derivations, $X_0, X_1, X_2, X_3, \ldots$, one for each "generation", such that for all $t \in \mathbb{N}$, $X_t \preceq X_{t+1}$. $X_0 = (B_0, s_0)$ is the initial derivation, at generation 0, and consists of a single (root) node $B_0 = \{\varepsilon\}$, labeled by the start non-terminal, $s_0(\varepsilon) = T_i$. Inductively, for all $t \in \mathbb{N}$ the derivation $X_{t+1} = (B_{t+1}, s_{t+1})$ is obtained from $X_t = (B_t, s_t)$ as follows. For each leaf $w \in \mathcal{L}_{B_t}$:

- if $s_t(w) = T_i$ is a Q-form (branching) non-terminal, whose associated unique rule is $T_i \xrightarrow{1} T_j \, T_{j'}$, then w must have exactly two children in B_{t+1}, namely $wl \in B_{t+1}$ and $wr \in B_{t+1}$, and moreover we must have $s_{t+1}(wl) = T_j$ and $s_{t+1}(wr) = T_{j'}$.
- if $s_t(w) = T_i$ is a L-form (probabilistic) non-terminal, then w has exactly one child in B_{t+1}, namely wu, and for each rule $T_i \xrightarrow{p_{i,j}} T_j$ with $p_{i,j} > 0$, the probability that $s_{t+1}(wu) = T_j$ is $p_{i,j}$, and likewise when $T_i \xrightarrow{p_{i,0}} \varnothing$, is a rule with $p_{i,0} > 0$, then $s_{t+1}(wu) = \varnothing$ with probability $p_{i,0}$.
- if $s_t(w) = T_i$ is a M-form (controlled) non-terminal, then w has exactly one child in B_{t+1}, namely wu, and for each action $a_z \in \Gamma^i$, with probability $\sigma(h_w)(a_z)$, $s_{t+1}(wu) = T_{j_z}$, where $T_i \xrightarrow{a_z} T_{j_z}$ is the rule associated with a_z.

There are no other nodes in B_{t+1}. In particular, if $s_t(w) = \varnothing$, then in B_{t+1} the node w has no children. This defines a stochastic process, $X_0, X_1, X_2, \ldots$, where $X_t \preceq X_{t+1}$, for all $t \in \mathbb{N}$, and such that there is a unique play, $X = \lim_{t \to \infty} X_t$, such that $X_t \preceq X$ for all $t \in \mathbb{N}$. In this sense, the random process defines a probability space of plays.

For our purposes, an *objective* is specified by a property (i.e., a measurable set), $\mathcal{F}$, of plays, whose probability the player wishes to optimize (maximize or minimize). For a property $\mathcal{F}$ and a strategy $\sigma \in \Psi$, let $Pr_{T_i}^{\sigma}[\mathcal{F}]$ denote the probability that starting at non-terminal T_i, under strategy σ, the generated play is in the set $\mathcal{F}$. Let $Pr_{T_i}^*[\mathcal{F}] := \sup_{\sigma \in \Psi} Pr_{T_i}^{\sigma}[\mathcal{F}]$. For a non-terminal T_q, $q \in [n]$, let $Reach(T_q)$ denote the set of plays that contain T_q as a label of some node. Let $Reach^{\complement}(T_q)$ denote the complement event, i.e., the set of plays that do not contain T_q. A rather general form of *quantitative* multi-objective reachability decision problems that one might wish to consider is whether there exists a strategy $\sigma' \in \Psi$ such that a boolean combination of statements of the form $Pr_{T_i}^{\sigma'}[\mathcal{F}_j] \triangle_j p_j$ holds, where $\triangle_j \in \{<, \leq, =, \geq, >\}$, and where $\mathcal{F}_j$ is itself a boolean combination (using union and intersection) of (non-)reachability objectives of the form $Reach(T_{j_k})$ and $Reach^C(T_{j_k})$.

Our primary focus is on the following two *qualitative* multi-objective reachability problems. Given an OBMDP, $\mathcal{A}$ with non-terminals $V = \{T_1, \ldots, T_n\}$, given a start non-terminal T_i, and given set $K \subseteq [n]$ of targets, we wish to decide:

- (almost-sure): does there exist $\sigma \in \Psi$ such that $\bigwedge_{q \in K} Pr_{T_i}^{\sigma}[Reach(T_q)] = 1$?
  (Equivalently, does there exist $\sigma \in \Psi$ s.t. $Pr_{T_i}^{\sigma}[\bigcap_{q \in K} Reach(T_q)] = 1$?[7])
- (limit-sure): Is there, for every $\epsilon > 0$, a $\sigma_\epsilon \in \Psi$, s.t. $\bigwedge_{q \in K} Pr_{T_i}^{\sigma_\epsilon}[Reach(T_q)] \geq 1 - \epsilon$? (Equivalently, is $Pr_{T_i}^*[\bigcap_{q \in K} Reach(T_q)] = 1$?)

As mentioned, when $|K| = 1$, the almost-sure and limit-sure questions are equivalent [8]. The following example shows this is not so when $|K| \geq 2$:

Example 1. Consider the OBMDP with non-terminals $\{M, A, R_1, R_2\}$, and with target non-terminals $\{R_1, R_2\}$. M is the only "controlled" non-terminal, and the rules are[8]:

$$M \xrightarrow{a} M A \qquad\qquad A \xrightarrow{1/2} R_1$$

$$M \xrightarrow{b} R_2 \qquad\qquad A \xrightarrow{1/2} \varnothing$$

The supremum probability, $Pr_M^*[Reach(R_1) \cap Reach(R_2)]$, starting with non-terminal M, of reaching both targets is 1. To see this, for any $\epsilon > 0$, let the strategy keep choosing deterministically the action a until $l := \lceil \log_2(\frac{1}{\epsilon}) \rceil$ copies of non-terminal A have been created. Then in the (unique) copy of non-terminal M in generation l the strategy switches deterministically to action b. The probability of reaching target R_2 is 1. The probability of reaching R_1 is $1 - 2^{-l} \geq 1 - \epsilon$. Hence $Pr_M^*[Reach(R_1) \cap Reach(R_2)] = 1$.

However, $\nexists \sigma \in \Psi : \ Pr_M^{\sigma}[Reach(R_1) \cap Reach(R_2)] = 1$. To see this, note that if the strategy ever puts positive probability on action b in any "round", then with positive probability target R_1 will not be reached in the play. So, to reach target R_1 with probability 1, the strategy must deterministically choose action a forever, from every occurrence of non-terminal M. But if it does this the probability of reaching target R_2 would be 0. □

We now observe (proof in the full version) that qualitative multi-objective reachability problems over an unbounded target set K are in general NP-hard.

---

[7] The fact that these statements are equivalent is easy to prove; see the full version.

[8] Technically, as given, this OBMDP in not in simple normal form; but this can easily be rectified by using an auxiliary branching non-terminal, Q, adding the rule $Q \xrightarrow{1} M A$ and changing the rule $M \xrightarrow{a} M A$ to $M \xrightarrow{a} Q$.

Proposition 1

(1.) *The following two problems are both NP-hard: given an OBMDP, a set $K \subseteq [n]$ of target non-terminals, and a start non-terminal $T_i \in V$, decide whether (i) $\exists \sigma \in \Psi : Pr^\sigma_{T_i}[\bigcap_{q \in K} Reach(T_q)] = 1$, & (ii) $Pr^*_{T_i}[\bigcap_{q \in K} Reach(T_q)] = 1$.*

(2.) *The following problem is coNP-hard: given an OBP (i.e., an OBMDP with no controlled non-terminals, and hence with only one trivial strategy σ), a set $K \subseteq [n]$ of target non-terminals, and a start non-terminal $T_i \in V$, decide whether $Pr^\sigma_{T_i}[\bigcap_{q \in K} Reach(T_q)] = 0$.*

The proof is a reduction from 3-SAT for (1.), and from its complement for (2.).

We shall hereafter often use the notation $T_i \to T_j$ (respectively, $T_i \not\to T_j$), to denote that for non-terminal T_i there exists (respectively, there does *not* exist) either an associated (controlled) rule $T_i \xrightarrow{a} T_j$, where $a \in \Gamma^i$, or an associated probabilistic rule $T_i \xrightarrow{p_{i,j}} T_j$ with positive probability $p_{i,j} > 0$. Similarly let $T_i \to \varnothing$ (respectively, $T_i \not\to \varnothing$), denote that the rule $T_i \xrightarrow{p_{i,0}} \varnothing$ has positive probability $p_{i,0} > 0$ (respectively, has probability $p_{i,0} = 0$).

Definition 2. *The dependency graph of an OBMDP, $\mathcal{A}$, is a directed graph that has a node T_i for each non-terminal T_i, and contains an edge (T_i, T_j) if and only if: either $T_i \to T_j$ or there is a rule $T_i \xrightarrow{1} T_j T_r$ or a rule $T_i \xrightarrow{1} T_r T_j$ in $\mathcal{A}$.*

For an OBMDP, $\mathcal{A}$, with non-terminals set V, we let $G = (U, E)$, with $U = V$, denote the dependency graph of $\mathcal{A}$, and we use $G[C]$ to denote the subgraph of G induced by the subset $C \subseteq U$ of nodes (non-terminals).

Definition 3. *For a directed graph $G = (U, E)$, given a partition of its vertices $U = (U_1, U_P)$, an end-component is a set of vertices $C \subseteq U$ such that $G[C]$: (1) is strongly connected; (2) for all $u \in U_P \cap C$ and all $(u, u') \in E$, $u' \in C$; (3) and if $C = \{u\}$ (i.e., $|C| = 1$), then $(u, u) \in E$. A maximal end-component (MEC) is an end-component not contained in any larger end-component. A MEC-decomposition is a partition of the graph into MECs and nodes that do not belong to any MEC.*

MECs are disjoint and the unique MEC-decomposition of such a directed graph G (with a given partition of its nodes) can be computed in P-time ([4]).[9] More recent work provides more efficient algorithms for computing a MEC-decomposition [2]. For an OBMDP dependency graph $G = (U, E)$, $U = V$, the partition of U we use is: $U_P := \{T_i \in U \mid T_i$ is of L-form$\}$ and $U_1 := \{T_i \in U \mid T_i$ is of M-form or Q-form$\}$. We will also be using the notion of a strongly connected component (SCC) of a dependency graph, which can be defined as a MEC where condition (2) from Definition 3 above is not required. As is well-known, an SCC-decomposition of a digraph can be computed in linear time.

---

[9] In [4], maximal end-components are referred to as *closed components*.

3   Algorithm for deciding $Pr^*_{T_i}[\bigcap_{q \in K} Reach(T_q)] \overset{?}{=} 1$

We first note that there is a (relatively easy) algorithm to compute, for every subset of the target non-terminals $K' \subseteq K$, the sets $Z_{K'} := \{T_i \in V \mid \forall \sigma \in \Psi : Pr^\sigma_{T_i}[\bigcap_{q \in K'} Reach(T_q)] = 0\}$ and $\bar{Z}_{K'} := V - Z_{K'}$. This can be computed, in time via a suitable "attractor set" construction and dynamic programming, using as an initialization step an algorithm from [8, Proposition 4.1] for the single-target case. (See the full version for the algorithm and proof.)

Proposition 2. *Given an OBMDP, $\mathcal{A}$, and a set $K \subseteq [n]$ of $k = |K|$ target non-terminals, there is an algorithm that computes, for every subset of target non-terminals $K' \subseteq K$, the set $Z_{K'} := \{T_i \in V \mid \forall \sigma \in \Psi : Pr^\sigma_{T_i}[\bigcap_{q \in K'} Reach(T_q)] = 0\}$. The algorithm runs in time $4^k \cdot |\mathcal{A}|^{O(1)}$. The algorithm can also be augmented to compute a deterministic (non-static) strategy $\sigma'_{K'}$ and a rational value $b_{K'} > 0$, such that for all $T_i \notin Z_{K'}$, $Pr^{\sigma'_{K'}}_{T_i}[\bigcap_{q \in K'} Reach(T_q)] \geq b_{K'} > 0$.*

We now present the algorithm for deciding limit-sure multi-target reachability, i.e., whether $Pr^*_{T_i}[\bigcap_{q \in K} Reach(T_q)] \doteq \sup_{\sigma \in \Psi} Pr^\sigma_{T_i}[\bigcap_{q \in K} Reach(T_q)] = 1$.

First, as a preprocessing step, for each subset of target non-terminals $K' \subseteq K$, we compute the set $Z_{K'} := \{T_i \in V \mid \forall \sigma \in \Psi : Pr^\sigma_{T_i}[\bigcap_{q \in K'} Reach(T_q)] = 0\}$, using the algorithm from Proposition 2. For every $q \in K$, let AS_q denote the set of non-terminals T_j (including T_q itself) such that $Pr^*_{T_j}[Reach(T_q)] = 1$. The set AS_q can be computed in P-time ([8, Theorem 9.3]), for each target non-terminal T_q, $q \in K$. Moreover, it was proved in [8, Theorem 9.4] that for (O)BMDPs the single-target almost-sure and limit-sure reachability problems coincide. So, for every $q \in K$, there exists a strategy τ_q such that $\forall T_j \in AS_q :$ $Pr^{\tau_q}_{T_j}[Reach(T_q)] = 1$. Let K'_{-i} denote the set $K' - \{i\}$.

Theorem 1. *The algorithm in Fig. 1 computes, given an OBMDP, $\mathcal{A}$, and a set $K \subseteq [n]$ of $k = |K|$ target non-terminals, for each subset $K' \subseteq K$, the set of non-terminals $F_{K'} := \{T_i \in V \mid Pr^*_{T_i}[\bigcap_{q \in K'} Reach(T_q)] = 1\}$. The algorithm runs in time $4^k \cdot |\mathcal{A}|^{O(1)}$. Moreover, for each $K' \subseteq K$, given $\epsilon > 0$, the algorithm can also be augmented to compute a randomized non-static strategy $\sigma^\epsilon_{K'}$ such that $Pr^{\sigma^\epsilon_{K'}}_{T_i}[\bigcap_{q \in K'} Reach(T_q)] \geq 1 - \epsilon$ for all non-terminals $T_i \in F_{K'}$.*

We omit the proof and instead provide some brief intuition for why the algorithm works. (The full proof also describes how the algorithm can be augmented to output, when given $\epsilon > 0$ as input, the witness strategy $\sigma^\epsilon_{K'}$.) For any subset $K' \subseteq K$ of target non-terminals, the set $D_{K'}$ contains the non-terminals starting from which, by induction using "smaller" target sets, it immediately follows that limit-sure multi-target reachability of K' holds. For instance, $D_{K'}$ contains any controlled (M-form) non-terminal T_i, $i \in K'$, with a rule $T_i \overset{a_j}{\to} T_j$ such that the remaining targets $K' - \{i\}$ can be limit-surely reached starting from T_j. The set $S_{K'}$ accumulates the non-terminals $T_i \in X = V - (D_{K'} \cup Z_{K'})$ such that there is a value $g > 0$ such that for any $\sigma \in \Psi : Pr^\sigma_{T_i}[\bigcap_{q \in K'} Reach(T_q)] \leq 1 - g$. In

I. Let $F_{\{q\}} := AS_q$, for each $q \in K$. $F_\emptyset := V$.

II. For $l = 2 \ldots k$:

For every subset of target non-terminals $K' \subseteq K$ of size $|K'| = l$:

1. $D_{K'} := \{T_i \in V - Z_{K'} \mid$ one of the following holds:
   - T_i is of L-form where $i \in K'$, $T_i \nrightarrow \emptyset$ and $\forall T_j \in V$: if $T_i \to T_j$, then $T_j \in F_{K'_{-i}}$.
   - T_i is of M-form where $i \in K'$ and $\exists a^* \in \Gamma^i : T_i \xrightarrow{a^*} T_j$, $T_j \in F_{K'_{-i}}$.
   - T_i is of Q-form $(T_i \xrightarrow{1} T_j \ T_r)$ where $i \in K'$ and $\exists K_L \subseteq K'_{-i} : T_j \in F_{K_L} \wedge T_r \in F_{K'_{-i} - K_L}$.
   - T_i is of Q-form $(T_i \xrightarrow{1} T_j \ T_r)$ where $\exists K_L \subset K' \ (K_L \neq \emptyset) : T_j \in F_{K_L} \wedge T_r \in F_{K' - K_L}.\}$

2. Repeat until no change has occurred to $D_{K'}$:
   (a) add $T_i \notin D_{K'}$ to $D_{K'}$, if of L-form, $T_i \nrightarrow \emptyset$ and $\forall T_j \in V$: if $T_i \to T_j$, then $T_j \in D_{K'}$.
   (b) add $T_i \notin D_{K'}$ to $D_{K'}$, if of M-form and $\exists a^* \in \Gamma^i : T_i \xrightarrow{a^*} T_j$, $T_j \in D_{K'}$.
   (c) add $T_i \notin D_{K'}$ to $D_{K'}$, if of Q-form $(T_i \xrightarrow{1} T_j \ T_r)$ and $T_j \in D_{K'} \vee T_r \in D_{K'}$.

3. Let $X := V - (D_{K'} \cup Z_{K'})$.

4. Initialize $S_{K'} := \{T_i \in X \mid$ either $i \in K'$, or T_i is of L-form and $T_i \to \emptyset \vee T_i \to T_j$, $T_j \in Z_{K'}\} \cup \bigcup_{\emptyset \subset K'' \subset K'} (X \cap S_{K''})$.

5. Repeat until no change has occurred to $S_{K'}$:
   (a) add $T_i \in X - S_{K'}$ to $S_{K'}$, if of L-form and $T_i \to T_j$, $T_j \in S_{K'} \cup Z_{K'}$.
   (b) add $T_i \in X - S_{K'}$ to $S_{K'}$, if of M-form and $\forall a \in \Gamma^i : T_i \xrightarrow{a} T_j$, $T_j \in S_{K'} \cup Z_{K'}$.
   (c) add $T_i \in X - S_{K'}$ to $S_{K'}$, if of Q-form $(T_i \xrightarrow{1} T_j \ T_r)$ and $T_j \in S_{K'} \cup Z_{K'} \wedge T_r \in S_{K'} \cup Z_{K'}$.

6. $\mathcal{C} \leftarrow$ MEC decomposition of $G[X - S_{K'}]$.

7. For every $q \in K'$, let $H_q := \{T_i \in X - S_{K'} \mid T_i$ is of Q-form $(T_i \xrightarrow{1} T_j \ T_r)$ and $((T_j \in X - S_{K'} \wedge T_r \in \bar{Z}_{\{q\}}) \vee (T_j \in \bar{Z}_{\{q\}} \wedge T_r \in X - S_{K'}))\}$.

8. Let $F_{K'} := \bigcup \{C \in \mathcal{C} \mid P_C = K' \vee (P_C \neq \emptyset \wedge P_C \neq K' \wedge \exists T_i \in C, \exists a \in \Gamma^i : T_i \xrightarrow{a} T_j, T_j \in F_{K' - P_C})\}$, where $P_C = \{q \in K' \mid C \cap H_q \neq \emptyset\}$.

9. Repeat until no change has occurred to $F_{K'}$:
   (a) add $T_i \in X - (S_{K'} \cup F_{K'})$ to $F_{K'}$, if of L-form and $T_i \to T_j$, $T_j \in F_{K'} \cup D_{K'}$.
   (b) add $T_i \in X - (S_{K'} \cup F_{K'})$ to $F_{K'}$, if of M-form and $\exists a^* \in \Gamma^i : T_i \xrightarrow{a^*} T_j$, $T_j \in F_{K'}$.
   (c) add $T_i \in X - (S_{K'} \cup F_{K'})$ to $F_{K'}$, if of Q-form $(T_i \xrightarrow{1} T_j \ T_r)$ and $T_j \in F_{K'} \vee T_r \in F_{K'}$.

10. If $X \neq S_{K'} \cup F_{K'}$, let $S_{K'} := X - F_{K'}$ and go to step 5.

11. Else, i.e., if $X = S_{K'} \cup F_{K'}$, let $F_{K'} := F_{K'} \cup D_{K'}$.

III. **Output** F_K.

Fig. 1. Algorithm for limit-sure multi-target reachability. The output is the set $F_K = \{T_i \in V \mid Pr^*_{T_i}[\bigcap_{q \in K} Reach(T_q)] = 1\}$.

other words, $S_{K'}$ will accumulate those non-terminals starting from which limit-sure reachability definitely does not hold. The loop in step II.5. is an attractor set construction that adds non-terminals T_i to set $S_{K'}$ based on prior membership in $S_{K'}$ of non-terminals appearing on the right-hand side of rules for non-terminal T_i. Step II.6. then builds a MEC-decomposition of the dependency graph $G[X - S_{K'}]$ induced by the remaining non-terminals in set $X - S_{K'}$; step II.8. identifies those MECs, C, where starting at a non-terminal in C the following is observed: the branching (Q-Form) non-terminals in C spawn two children each, at least one of which belongs to C, and other spawned children of the branching non-terminals in C can collectively reach a non-empty subset P_C of (or in the best case, all of) the target set K' with a positive probability (bounded away from zero); the player can choose to delay arbitrarily long the moment to select an action that "exits" C and, thus, can choose to reach the targets in set P_C with probability arbitrarily close to 1; and once the player chooses to "exit" C, it does so in a non-terminal that can limit-surely reach the rest of the targets in set $K' - P_C$. Step II.9. accumulates in the set $F_{K'}$ the set of non-terminals that can almost-surely reach the set $D_{K'}$ or one of the MECs computed in step II.8. A key assertion is this: if in step II.11. we find all non-terminals from the set X are already either in set $S_{K'}$ or in set $F_{K'}$, then we are done: $F_{K'} \cup D_{K'}$ must constitute the set of all non-terminals starting in which the player can force limit-sure reachability of all targets in set K' in the same play; otherwise, all non-terminals in set $X - (F_{K'} \cup S_{K'})$ can be added to set $S_{K'}$, meaning that starting at any of these non-terminals, limit-sure reachability of all target non-terminals in set K' cannot be achieved. This latter assertion is not obvious, but it is true (see the proof in the full version).

4   Algorithm for deciding whether $\exists \sigma \in \Psi$: $Pr_{T_i}^{\sigma}[\bigcap_{q \in K} Reach(T_q)] = 1$

We now present the algorithm (Fig. 2) for deciding almost-sure multi-target reachability for a given OBMDP, $\mathcal{A}$, i.e., given a set $K \subseteq [n]$ of $k = |K|$ target non-terminals and a starting non-terminal T_i, deciding whether there is a strategy for the player under which the probability of generating a play that contains all target non-terminals from set K is 1. Again, as in the limit-sure algorithm, for each subset of target non-terminals $K' \subseteq K$, as a preprocessing step we compute the set $Z_{K'} := \{T_i \in V \mid \forall \sigma \in \Psi : Pr_{T_i}^{\sigma}[\bigcap_{q \in K'} Reach(T_q)] = 0\}$. And for every $q \in K$, we compute (in P-time) the set AS_q of non-terminals T_j (including target T_q itself) such that there is a strategy τ with $Pr_{T_j}^{\tau}[Reach(T_q)] = 1$.

Theorem 2. *The algorithm in Fig. 2 computes, given an OBMDP, $\mathcal{A}$, and a set $K \subseteq [n]$ of $k = |K|$ target non-terminals, for each subset $K' \subseteq K$, the set of non-terminals $F_{K'} := \{T_i \in V \mid \exists \sigma \in \Psi : Pr_{T_i}^{\sigma}[\bigcap_{q \in K'} Reach(T_q)] = 1\}$. The algorithm runs in time $4^k \cdot |\mathcal{A}|^{O(1)}$. Moreover, for each $K' \subseteq K$, the algorithm can also be augmented to compute a randomized non-static strategy $\sigma_{K'}^*$ such that $Pr_{T_i}^{\sigma_{K'}^*}[\bigcap_{q \in K'} Reach(T_q)] = 1$ for all non-terminals $T_i \in F_{K'}$.*

I. Let $F_{\{q\}} := AS_q$, for each $q \in K$. $F_\emptyset := V$.

II. For $l = 2 \ldots k$:

For every subset of target non-terminals $K' \subseteq K$ of size $|K'| = l$:

1. $D_{K'} := \{T_i \in V - Z_{K'} \mid$ one of the following holds:
   - T_i is of L-form where $i \in K'$, $T_i \not\to \emptyset$ and $\forall T_j \in V$: if $T_i \to T_j$, then $T_j \in F_{K'_{-i}}$.
   - T_i is of M-form where $i \in K'$ and $\exists a^* \in \Gamma^i : T_i \xrightarrow{a^*} T_j$, $T_j \in F_{K'_{-i}}$.
   - T_i is of Q-form $(T_i \xrightarrow{1} T_j\, T_r)$ where $i \in K'$ and $\exists K_L \subseteq K'_{-i} : T_j \in F_{K_L} \wedge T_r \in F_{K'_{-i} - K_L}$.
   - T_i is of Q-form $(T_i \xrightarrow{1} T_j\, T_r)$ where $\exists K_L \subset K'\ (K_L \neq \emptyset) : T_j \in F_{K_L} \wedge T_r \in F_{K' - K_L}.\}$

2. Repeat until no change has occurred to $D_{K'}$:
   (a) add $T_i \notin D_{K'}$ to $D_{K'}$, if of L-form, $T_i \not\to \emptyset$ and $\forall T_j \in V$: if $T_i \to T_j$, then $T_j \in D_{K'}$.
   (b) add $T_i \notin D_{K'}$ to $D_{K'}$, if of M-form and $\exists a^* \in \Gamma^i : T_i \xrightarrow{a^*} T_j$, $T_j \in D_{K'}$.
   (c) add $T_i \notin D_{K'}$ to $D_{K'}$, if of Q-form $(T_i \xrightarrow{1} T_j\, T_r)$ and $T_j \in D_{K'} \vee T_r \in D_{K'}$.

3. Let $X := V - (D_{K'} \cup Z_{K'})$.

4. Initialize $S_{K'} := \{T_i \in X \mid$ either $i \in K'$, or T_i is of L-form and $T_i \to \emptyset \vee T_i \to T_j$, $T_j \in Z_{K'}\} \cup \bigcup_{\emptyset \subset K'' \subset K'} (X \cap S_{K''})$.

5. Repeat until no change has occurred to $S_{K'}$:
   (a) add $T_i \in X - S_{K'}$ to $S_{K'}$, if of L-form and $T_i \to T_j$, $T_j \in S_{K'} \cup Z_{K'}$.
   (b) add $T_i \in X - S_{K'}$ to $S_{K'}$, if of M-form and $\forall a \in \Gamma^i : T_i \xrightarrow{a} T_j$, $T_j \in S_{K'} \cup Z_{K'}$.
   (c) add $T_i \in X - S_{K'}$ to $S_{K'}$, if of Q-form $(T_i \xrightarrow{1} T_j\, T_r)$ and $T_j \in S_{K'} \cup Z_{K'} \wedge T_r \in S_{K'} \cup Z_{K'}$.

6. $\mathcal{C} \leftarrow$ SCC decomposition of $G[X - S_{K'}]$.

7. For every $q \in K'$, let $H_q := \{T_i \in X - S_{K'} \mid T_i$ is of Q-form $(T_i \xrightarrow{1} T_j\, T_r)$ and $((T_j \in X - S_{K'} \wedge T_r \in \bar{Z}_{\{q\}}) \vee (T_j \in \bar{Z}_{\{q\}} \wedge T_r \in X - S_{K'}))\}$.

8. Let $F_{K'} := \bigcup \{\bigcup_{q \in K'} (H_q \cap C) \mid C \in \mathcal{C}$ s.t. $\forall q' \in K' : H_{q'} \cap C \neq \emptyset\}$.

9. Repeat until no change has occurred to $F_{K'}$:
   (a) add $T_i \in X - (S_{K'} \cup F_{K'})$ to $F_{K'}$, if of L-form and $T_i \to T_j$, $T_j \in F_{K'} \cup D_{K'}$.
   (b) add $T_i \in X - (S_{K'} \cup F_{K'})$ to $F_{K'}$, if of M-form and $\exists a^* \in \Gamma^i : T_i \xrightarrow{a^*} T_j$, $T_j \in F_{K'}$.
   (c) add $T_i \in X - (S_{K'} \cup F_{K'})$ to $F_{K'}$, if of Q-form $(T_i \xrightarrow{1} T_j\, T_r)$ and $T_j \in F_{K'} \vee T_r \in F_{K'}$.

10. If $X \neq S_{K'} \cup F_{K'}$, let $S_{K'} := X - F_{K'}$ and go to step 5.

11. Else, i.e., if $X = S_{K'} \cup F_{K'}$, let $F_{K'} := F_{K'} \cup D_{K'}$.

III. **Output** F_K.

Fig. 2. Algorithm for almost-sure multi-target reachability. The output is the set $F_K = \{T_i \in V \mid \exists \sigma \in \Psi : Pr^\sigma_{T_i}[\bigcap_{q \in K} Reach(T_q)] = 1\}$.

We again omit the proof and instead provide a brief sketch of why the algorithm works. (The full proof also describes how the algorithm can be augmented to output the witness strategy $\sigma_{K'}^*$.) Both the sketch and the algorithm itself are very similar to that of the limit-sure case, but differ in some crucial details. Not only do the two algorithms differ in steps II.6. and II.8., but moreover the interpretation of various sets being accumulated in the two algorithms changes (in order to correspond to the appropriate meaning in the context of almost-sure reachability). For any subset $K' \subseteq K$ of target non-terminals, the set $D_{K'}$ contains the non-terminals starting from which, by induction using "smaller" target sets, it immediately follows that almost-sure multi-target reachability is satisfied. The set $S_{K'}$ accumulates the non-terminals $T_i \in X = V - (D_{K'} \cup Z_{K'})$ such that $\forall \sigma \in \Psi : Pr_{T_i}^\sigma [\bigcap_{q \in K'} Reach(T_q)] < 1$. In other words, $S_{K'}$ will accumulate those non-terminals starting from which almost-sure reachability definitely does not hold. The loop in step II.5. is again an attractor set construction that adds non-terminals T_i to set $S_{K'}$ based on prior membership in $S_{K'}$ of non-terminals appearing on the right-hand side of rules for non-terminal T_i. Step II.6. then builds a SCC-decomposition of the dependency graph $G[X - S_{K'}]$ induced by the remaining non-terminals in set $X - S_{K'}$; step II.8. identifies those branching (Q-form) non-terminals that belong to SCCs, C, where the following is true for each such C: the Q-form non-terminals in C (that have been identified in step II.8.) spawn two children each, at least one of which belongs to C, and the other spawned children of these same branching non-terminals can collectively reach all the targets in set K' with a positive probability (bounded away from zero).[10] Step II.9. accumulates in the set $F_{K'}$ the set of non-terminals that can almost-surely reach the set $D_{K'}$ or the Q-form non-terminals computed in step II.8. A key assertion is this: if in step II.11. we find all non-terminals from the set X are already either in set $S_{K'}$ or in set $F_{K'}$, then we are done: $F_{K'} \cup D_{K'}$ must constitute the set of all non-terminals starting in which the player can force almost-sure reachability of all targets in set K' in the play[11]; otherwise, all non-terminals in set $X - (F_{K'} \cup S_{K'})$ can be added to set $S_{K'}$, meaning that starting at any of these non-terminals, almost-sure reachability of all target

---

[10] Note that this differs crucially from the situation in the limit-sure algorithm, where the other spawned children of these branching nodes in C were only able to reach a non-empty *subset* P_C of K' with a positive probability (bounded away from zero), not necessarily the entire set K'.

[11] A helpful observation here is this: in the limit-sure algorithm we were identifying MECs, where the choice of when to "exit" the MEC is entirely controlled by the player. In the almost-sure algorithm we instead identify SCCs. Even though there may also be purely probabilistic (i.e., not controlled) opportunities of "exiting" such a "good" SCC, C (specifically, an SCC C that is identified and used in step II.8.), due to the way the set $F_{K'} = X - S_{K'}$ is constructed (and due to properties of its associated witness strategy $\sigma_{K'}^*$, which is described in the full proof), we can show that even when C is "exited" we still stay inside the set $F_{K'}$, and eventually hit a SCC, C', which can only be "exited" probabilistically to $D_{K'}$, and where, moreover, for each target in K' there is a branching (Q-Form) node in the SCC, C', whose "extra" child can hit that target with positive probability (bounded away from zero).

non-terminals in set K' cannot be achieved. The reason why this last assertion holds is again not obvious, but it is true (see the proof in the full version).

References

1. Bozic, I., et al.: Evolutionary dynamics of cancer in response to targeted combination therapy. eLife **2**, e00747 (2013)
2. Chatterjee, K., Henzinger, M.: Efficient and dynamic algorithms for alternating Büchi games and maximal end-component decomposition. J. ACM **61**(3), 15:1–15:40 (2014)
3. Chen, T., Dräger, K., Kiefer, S.: Model checking stochastic branching processes. In: Rovan, B., Sassone, V., Widmayer, P. (eds.) MFCS 2012. LNCS, vol. 7464, pp. 271–282. Springer, Heidelberg (2012). https://doi.org/10.1007/978-3-642-32589-2_26
4. Courcoubetis, C., Yannakakis, M.: Markov decision processes and regular events. IEEE Trans. Autom. Control **43**(10), 1399–1418 (1998)
5. Durbin, R., Eddy, S.R., Krogh, A., Mitchison, G.: Biological Sequence Analysis: Probabilistic Models of Proteins and Nucleic Acids. Cambridge University Press, Cambridge (1998)
6. Etessami, K., Stewart, A., Yannakakis, M.: A Polynomial-time algorithm for computing extinction probabilities of multi-type branching processes. SIAM J. Computing **46**(5), 1515–1553 (2017). (Incorporates part of the work of a conference paper in STOC 2012.)
7. Etessami, K., Stewart, A., Yannakakis, M.: Polynomial time algorithms for branching Markov decision processes and probabilistic min(max) polynomial Bellman equations. Math. Oper. Res. **45**(1), 34–62 (2020). (Conference version in ICALP'12.)
8. Etessami, K., Stewart, A., Yannakakis, M.: Greatest fixed points of probabilistic min/max polynomial equations, and reachability for branching Markov decision processes. Inf. Comput. **261**(2), 355–382 (2018). (Conference version in ICALP 2015.)
9. Etessami, K., Martinov, E., Stewart, A., Yannakakis, M.: Reachability for branching concurrent stochastic games. In: Proceedings of the 46th International Colloquium on Automata, Languages and Programming (ICALP) (2019). (All references are to the full preprint arXiv:1806.03907.)
10. Etessami, K., Martinov, E.: Qualitative multi-objective reachability for ordered branching MDPs. Full preprint arXiv:2008.10591 (2020)
11. Etessami, K., Kwiatkowska, M., Vardi, M.Y., Yannakakis, M.: Multi-objective model checking of Markov decision processes. LMCS **4**(4), (2008)
12. Etessami, K., Yannakakis, M.: Recursive concurrent stochastic games. LMCS **4**(4), (2008)
13. Etessami, K., Yannakakis, M.: Recursive Markov decision processes and recursive stochastic games. J. ACM **62**(2), 1–69 (2015)
14. Etessami, K., Yannakakis, M.: Recursive Markov chains, stochastic grammars, and monotone systems of nonlinear equations. J. ACM **56**(1), 1–66 (2009)
15. Haccou, P., Jagers, P., Vatutin, V.A.: Branching Processes: Variation, Growth, and Extinction of Populations. Cambridge University Press, Cambridge (2005)
16. Kimmel, M., Axelrod, D.E.: Branching Processes in Biology. IAM, vol. 19. Springer, New York (2002). https://doi.org/10.1007/b97371

17. Michalewski, H., Mio, M.: On the problem of computing the probability of regular sets of trees. In: Proceedings of FSTTCS 2015, pp. 489–502 (2015)
18. Przybyłko, M., Skrzypczak, M.: On the complexity of branching games with regular conditions. In: Proceedings of MFCS 2016, LIPICS, vol. 78 (2016)
19. Reiter, J.G., Bozic, I., Chatterjee, K., Nowak, M.A.: TTP: tool for tumor progression. In: Sharygina, N., Veith, H. (eds.) CAV 2013. LNCS, vol. 8044, pp. 101–106. Springer, Heidelberg (2013). https://doi.org/10.1007/978-3-642-39799-8_6

Quantum-over-Classical Advantage in Solving Multiplayer Games

Dmitry Kravchenko[1], Kamil Khadiev[2,3(✉)], Danil Serov[3], and Ruslan Kapralov[3]

[1] Center for Quantum Computer Science, Faculty of Computing, University of Latvia, Riga, Latvia
kravchenko@gmail.com
[2] Smart Quantum Technologies Ltd., Kazan, Russia
kamilhadi@gmail.com
[3] Kazan Federal University, Kazan, Russia
serovdanilru@gmail.com, kapralov_ruslan@mail.ru

Abstract. We study the applicability of quantum algorithms in computational game theory and generalize some results related to Subtraction games, which are sometimes referred to as one-heap Nim games.

In quantum game theory, a subset of Subtraction games became the first explicitly defined class of zero-sum combinatorial games with provable separation between quantum and classical complexity of solving them. For a narrower subset of Subtraction games, an exact quantum sublinear algorithm is known that surpasses all deterministic algorithms for finding solutions with probability 1.

Typically, both Nim and Subtraction games are defined for only two players. We extend some known results to games for three or more players, while maintaining the same classical and quantum complexities: $\Theta\left(n^2\right)$ and $\tilde{O}\left(n^{1.5}\right)$ respectively.

Keywords: Quantum game theory · Quantum combinatorial games · Quantum multiplayer games · Quantum algorithm · Nim · Subtraction game

1 Introduction

The concept of quantum games has been developed since the late 1990s [13,29,36]. Many examples of them have been invented to illustrate how different are the circumstances of the quantum microcosm from the "classical" ones.

Currently, theoretical and experimental results in the intersection of game theory and quantum computing can be divided into three groups, depending on how much the rules of a game relate to quantum effects.

Games with Quantum Decisions
These games are essentially quantum in the sense that they can be implemented only *quantum physically*.

© Springer Nature Switzerland AG 2020
S. Schmitz and I. Potapov (Eds.): RP 2020, LNCS 12448, pp. 83–98, 2020.
https://doi.org/10.1007/978-3-030-61739-4_6

Vaidman [36] was probably the first to speak about quantum protocol in terms of games. He illustrated the Greenberger-Horne-Zeilinger (GHZ) proof of the nonexistence of local hidden variables by presenting a 3-player game that cannot be won with confidence without using certain properties of quantum particles.

Meyer [29] presented an "unfair" coin-flipping game where one player is limited with "classical" strategies – "to flip the quantum coin" or "not to flip"— while the other player can apply any operation on the qubit. No wonder the latter always wins the game.

Eisert et al. [13] described a quantum version of Prisoner's Dilemma, where the prisoners have a better Nash equilibrium in quantum strategies. This work was followed by a series of criticism regarding the unreasonable limitations on the set of available quantum strategies, the inconsistency of the derived Nash equilibrium [6], and the concept of "detangling" the selected quantum strategies [14]. Nevertheless, this model of quantizing games was further developed into many directions [23,27], and many results are known about the existence and properties of Nash equilibria in games with quantum decisions. We refer to works [21,22,32,33,39] for the extensive studies of this kind of quantum games.

Nonlocal Games

The rules of nonlocal games do not rely in any way on any property of quantum particles. Such games are completely available for playing without any quantum technologies. However, exploiting certain quantum correlation properties results in a bigger probability of winning, compared to the best classical strategy. We note that in most cases the comparison is made regarding the *correlated Nash equilibria*, and the difference in results is always due to the difference between classical and quantum correlations.

Generally speaking, a nonlocal game is a cooperative game in which two or more players play against a referee. The referee randomly selects one question (from some predefined set of questions Q) for each player. After that, each player has to respond with an answer (from some predefined set of answers A). Players are allowed to communicate before the game started. After that, they are prohibited from transmitting any information to each other during the play of the game. Players win or lose according to predefined rules of the game.

The most famous example of a nonlocal game is, of course, CHSH [10], which can be won with probability at most 75% by classical players, or with probability 85.355... % when using a pair of entangled quantum particles. Several authors have shown a similar relationship between the game values for the quantum and classical versions of various nonlocal games for different numbers of players (value of a game is defined as difference between winning probability and losing probability) [3,4,37,38]. Besides, there were found games where this ratio reaches its maximum [5,28].

There are also several examples of not fully cooperative games and even of zero-sum games where quantum nonlocality brings an advantage for the players [9,24,30].

Quantum Algorithms for Solving Classical Games

This is the newest and perhaps the least developed group of "quantum games" in our list. We call them *quantum combinatorial games*, although they are essentially classical: neither the rules of these games nor the strategies assume any use of quantum technologies. However, sometimes quantum algorithms can help in finding good classical strategies.

We still do not know whether quantum computers are generally able to efficiently find Nash equilibria for classical games [25]. However, this assumption is indirectly corroborated by advances in quantum combinatorial optimization. More precisely, by the *quantum annealing* technique—including the one for solving combinatorial games [34].

This paper is devoted to yet another case of using quantum algorithms for that purpose. Hereafter we speak only about quantum combinatorial games.

First explicit examples of combinatorial games with quantum-better-than-classical solving algorithms are some subsets of *Subtraction games*. Khadiev, Kravchenko and Serov [20] identify a specific subset of size $const^{n^2}$ of Subtraction games which are solvable by a quantum algorithm in time $O\left(n^{1.5}\log n\right)$ in bounded error setting. Huang, Ye, Zheng and Li [18] identify a smaller set of size $const^{\sqrt{n}\log n}$ of restricted Subtraction games which are solvable by an exact quantum algorithm in time $O\left(n^{1.5}\right)$. Deterministic algorithms for both classes of games require $\Omega\left(n^2\right)$ steps for solving. (Hereafter $n+1$ stands for the number of positions in a game, regardless which of the players has to make the next move. *const* here stands for some not too big numbers greater than 1; their exact values may depend on the details of definitions.)

A *Subtraction game* is similar to a canonical Nim game [15] in several senses. Nim is a notable game in game theory because it traditionally serves as a "base case" for Sprague-Grundy theorem [17,35], which establishes a deterministic upper bound for solving many combinatorial games. Similarly, Subtraction games seem to become a good candidate for being a "base sample" for game-solving Quantum Dynamic Programming. As many games are known to be reducible to some Nim games, many games on graphs can be reduced to the corresponding Subtraction games. Finally, the rules of these games have very similar definitions.

The difference between these two games is that in a Subtraction game, the players deal with just one heap of stones, though with certain limitations imposed on the number of stones they can take from the heap. The most common limitation for Subtraction games is defining a maximum for the number of stones to be taken away. This kind of Subtraction game has very fast deterministic solutions in $\tilde{O}\left(1\right)$. Here we study a much more general class of such limitations and thus a broader class of Subtraction games.

We investigate algorithms for *solving* Subtraction games, which determine the payoffs of all the players, assuming each of them to play optimally. We exploit ideas similar to ones in [20] and [18] to establish upper bounds for the quantum complexity and asymptotes for the classical complexity. Surprisingly, some of these approaches fit better and are more natural for multiplayer games than for two-player games.

The paper is organized in the following way. Section 2 contains basic definitions. In Sect. 3, we present evaluations of classical complexity and quantum algorithms for solving a special class of games, which we call *balanced* Subtraction games. Finally, in Sect. 4 we analyze the complexity of solving the so-called *restricted* Subtraction games.

2 Definitions

2.1 Subtraction Games

In a play of a *Subtraction game* players $1, \ldots, k$ sequentially remove some positive amounts of stones from a heap, with player l being followed by player $(l \bmod k + 1)$.

Let n be the initial number of stones in the heap, and Γ be a lower-triangular *binary* matrix of size $n \times n$, with rows numbered from 1 to n and columns numbered from 0 to $n - 1$. A player which has to make the next move, can remove $j - i$ stones $(0 \leq i < j \leq n)$ from the heap with exactly j stones left iff $\Gamma_{ji} = 1$. In simple terms, Γ_{ji} indicates the possibility for a player to receive position "j stones" from the predecessor and pass position "i stones" to the follower on the next turn.

If in some position a player, say player l_0, cannot make a legal move, then the play ends, and each player l receives their payoff $(l - l_0 + k) \bmod k$. That is, player l_0 is the loser, the previous player is the major winner with payoff $(k - 1)$, the previous-to-previous player gets $(k - 2)$ and so on. In order to become the major winner, a player has to take all the remaining stones, or to leave a number of stones j such that no allowed moves would remain: $\sum_{i=0}^{j-1} \Gamma_{ji} = 0$.

Obviously, the rules of a Subtraction game are fully determined by such matrix Γ, so hereafter we use letter Γ to denote a corresponding game. We also reserve the name n to denote the initial number of stones. This number n also corresponds to the dimension of the matrix Γ. Note that there are only $n(n + 1)/2$ meaningful bits in the matrix Γ, as a player cannot increase the number of stones in the heap or leave it as is: $\Gamma_{ji} = 0$ for all $j \leq i$.

Finally, we note that the selected payoff function is not a must. It may be arbitrary, provided that each player has strict preferences over the set of all k possible endings. Otherwise, if some preferences are not strict, the concept of optimal behavior will not be well-defined, and the required assumption of optimal players will fail.

2.2 Payoff Function

Let $\mathcal{G}$ be a set of lower-triangular binary matrices of size $n \times n$, with rows numbered from 1 to n and columns numbered from 0 to $n - 1$.

We define a payoff function $\text{WIN} : \{(\Gamma, j) \,|\, \Gamma \in \mathcal{G} \,\wedge\, j \in \mathbb{N}_n\} \to \{w\}_{0 \leq w < k}$, such that $\text{WIN}(\Gamma, j) = w$ iff a player gets payoff w given position "j stones" in game Γ, under assumption of optimal players[1]:

$$\text{WIN}(\Gamma, j) = \begin{cases} 0 & \text{if } j = 0, \\ 0 & \text{if } \sum_i \Gamma_{ji} = 0, \\ \left(\max_{i:\Gamma_{ji}=1} \text{WIN}(\Gamma, i) - 1 \right) \bmod k & \text{otherwise.} \end{cases}$$

We also use notation $\text{WIN}(\Gamma) = \text{WIN}(\Gamma, n)$ for the value of game Γ.

Informally:

$\star$ a player loses if s/he receives the position "0 stones";
$\star$ a player loses if s/he receives a position "j stones" which does not suggest any legal move, according to the rules of the game;
$\star$ otherwise, a rational player wants to pass a losing position to their successor (the next player), so as to get the biggest possible payoff $k - 1$.
$\star$ if there is no such move, s/he instead passes the most preferable position to their successor and thus gets the payoff of the successor minus 1.

Note that for a certain position, rules of the game prescribe the same set of available moves for all players regardless of their numbers, hence $\text{WIN}(\Gamma, j)$ describes the final payoff for whoever receives position "j stones".

2.3 Example

Let us consider a 3-player Subtraction game where the initial number of stones is 7, and a player on their turn should remove some number of stones which is a power of 2. The corresponding game matrix is equal to

$$\Gamma = \begin{pmatrix} 1 & 0 & 0 & 0 & 0 & 0 & 0 \\ 1 & 1 & 0 & 0 & 0 & 0 & 0 \\ 0 & 1 & 1 & 0 & 0 & 0 & 0 \\ 1 & 0 & 1 & 1 & 0 & 0 & 0 \\ 0 & 1 & 0 & 1 & 1 & 0 & 0 \\ 0 & 0 & 1 & 0 & 1 & 1 & 0 \\ 0 & 0 & 0 & 1 & 0 & 1 & 1 \end{pmatrix}.$$

The values of positions $\text{WIN}(\Gamma, j)$ can be determined by a straightforward dynamic programming:

$\star$ "0 stones" is a losing position: $\text{WIN}(\Gamma, 0) = 0$;
$\star$ positions "1 stone", "2 stones", and "4 stones" allow a player to pass a losing position "0 stones" to the successor immediately by removing all the remaining stone(s), as $1, 2, 4$ are the powers of 2: $\text{WIN}(\Gamma, 1) = \text{WIN}(\Gamma, 2) = \text{WIN}(\Gamma, 4) = 2$;
$\star$ positions "3 stone", "5 stones", and "6 stones" do not allow to win 2, but it is possible there to pass a good position (e.g.. "1 stone" or "2 stones") to the successor and thus get payoff 1: $\text{WIN}(\Gamma, 3) = \text{WIN}(\Gamma, 5) = \text{WIN}(\Gamma, 6) = 1$;

---

[1] Note that in accordance with the chosen numbering of rows and columns of Γ, $\Gamma_{ji} = 1 \implies i < j$, so this recursive definition of the function WIN is valid.

$\star$ given position "7 stones", a player has no other option but to leave 3, 5 or 6 stones for their successor, so the successor gets payoff 1 and hence $\text{WIN}(\Gamma) = \text{WIN}(\Gamma, 7) = 0$.

It is not hard to solve this kind of game for bigger values of n and k, by properly analyzing combinatorial properties of removing 2^p stones, but in this work we consider Subtraction games with somewhat arbitrary rules which generally can differ for different positions.

2.4 Properties of Subtraction Games

In this work we stick to the conventional terminology of [20, Section 2.3] and [18, Section 2.1], and use the following definitions.

We call a game Γ *losing* if the first player loses it assuming other players are optimal:

$$\text{WIN}(\Gamma) = 0. \tag{1}$$

We call a game Γ *balanced* if the values of $\text{WIN}(\Gamma, j)$ are uniformly distributed over the set $\{w\}_{0 \leq w < k}$[2]:

$$\forall w : \left| \#\{j : \text{WIN}(\Gamma, j) = w\}_{1 \leq j \leq n} - \frac{n}{k} \right| \leq o\left(\frac{n}{k}\right). \tag{2}$$

When considering a random balanced game we hereafter implicitly bear in mind the following procedure of picking a game:

1. Assign each position "j stones" one of the values from $\{0, 1, \ldots, (k-1)\}$ in the uniform fashion.
2. Assign position "0 stones" value 0.
3. Initially assign $\Gamma = [0]_{ji}$.
4. For each position "j stones" with value w put $\Gamma_{ji} = 1$ with probability $1/2$ whenever $i < j$ and position "i stones" is assigned one of the values $(w+1) \bmod k$, w, $(w-1)$, $\ldots$, 1.
5. Additionally, for each position "j stones" with value w put $\Gamma_{ji} = 1$ for one i such that position "i stones" is assigned value $(w+1) \bmod k$, whenever it was not already done in the previous step.
6. If the previous step failed because for some position "j stones" it is not possible to find a position "i stones" with the appropriate value, then start everything from the beginning.[3]

---

[2] The term *balanced* naturally comes from the notion of *balanced functions*: balanced game Γ is such that its payoff function $\text{WIN}(\Gamma, j)$ is balanced. Formally, the definition of *perfect balancedness* should look like $\forall w : \#\{j : \text{WIN}(\Gamma, j) = w\}_{1 \leq j \leq n} = \frac{n}{k}$, but we use the little-$o$ notation to extend our results also to *almost balanced* games.

[3] Should one feel that discarding in this step essentially destroys the uniformity of $\text{WIN}(\Gamma, j)$, they can at step 2 assign each position "w stones", $0 \leq w < k$, value $(k-w) \bmod k$. This will make the last step obsolete, as no failure can occur, and will preserve the perfect uniformity. Our further observations are valid for either kind of picking a random balanced Subtraction game.

Finally, we call a game Γ *restricted* if in each position at most one move is possible:

$$\forall j : \sum_i \Gamma_{ji} \leq 1. \tag{3}$$

The sample game from the previous subsection is losing and balanced (as far as the defined notion of balancedness is applicable to $n = 7$), but not restricted.

2.5 Quantum Computing and Computational Model

Basics of Quantum Computing. The main difference between quantum computation and classical one is manipulations with quantum bits (qubits). A state of a qubit is a vector from 2-dimensional complex Hilbert space[4]. We can represent it using Dirac notation as $|\psi\rangle = a|0\rangle + b|1\rangle$, where $|0\rangle$ and $|1\rangle$ are unit vectors, and a and b are complex numbers such that $|a|^2 + |b|^2 = 1$. We can use two kind of transformations: *transition* and *measurement*. The transition is multiplying a vector of state to 2×2 unitary matrix. The measurement is obtaining 0-result with probability $|a|^2$ and 1-result with probability $|b|^2$. Similarly, a state of a register of q qubits is a vector from 2^q-dimensional complex Hilbert space, and is traditionally denoted as $|\psi\rangle = \sum_{i=0}^{2^q-1} a_i|i\rangle$, where $\sum_{i=0}^{2^q-1} |a_i|^2 = 1$. Transformations are defined in the analogous manner.

Computational Model. To evaluate the complexity of a quantum algorithm, we use the standard form of the quantum query model. It is a generalization of the decision tree model of classical computation that is commonly used to lower-bound the amount of time required by a computation.

Let $f : D \to \{0,1\}, D \subseteq \{0,1\}^M$ be an n variable function we wish to compute on an input $x \in D$. We have an oracle access to the input x, which is realized by a specific unitary transformation usually defined as $|i\rangle|z\rangle|w\rangle \to |i\rangle|z \oplus x_i\rangle|w\rangle$ where the $|i\rangle$ register indicates the index of the variable we are querying, $|z\rangle$ is the output register, and $|w\rangle$ is some auxiliary work-space. An algorithm in the query model consists of alternating applications of arbitrary unitaries independent of the input and the query unitary, and a measurement in the end. The smallest number of queries for an algorithm that outputs $f(x)$ with probability $\geq \frac{2}{3}$ on all x is called the quantum query complexity of the function f and is denoted by $Q(f)$. In this paper, as running time of an algorithm, we mean a number of queries to oracle. More information on quantum computation and query model can be found in [1,2,31].

To distinguish ordinary deterministic and randomized complexities from the quantum complexity, they are traditionally called by one term *classical complexity*.

---

[4] Formally, these vectors in the complex Hilbert space represent equivalence classes of vectors under multiplication by non-zero complex number. We also note that, for the purposes of this paper and throughout all the algorithms which we mention and refer here, one may assume all complex values to be in $\mathbb{R}$. However, in other important quantum algorithms the imaginary parts may play an essential role.

Grover's Algorithm for Quantum Search

Definition 1 (Search problem). *Suppose we have a set of objects named* $\{1, 2, \ldots, M\}$, *of which some are targets. Suppose* $\mathcal{O}$ *is an oracle that identifies the targets. The goal of a search problem is to find a target* $i \in \{1, 2, \ldots, M\}$ *by making queries to the oracle* $\mathcal{O}$.

In search problems, one will try to minimize the number of queries to the oracle. In the classical setting, one needs $O(M)$ queries to solve such a problem. Grover, on the other hand, constructed a quantum algorithm that solves the search problem with only $O(\sqrt{M})$ queries [16], provided that there is a unique target. For the case where the number of targets is arbitrary, say T, Brassard *et al.* designed a modified Grover algorithm that solves the search problem in $O(\sqrt{M/T})$ queries [8].

Quantum Algorithm for Maximum Search Problem

Definition 2 (Maximum search problem). *For some positive integers* D, L, R, *given a function* $f : \{1, \ldots, M\} \rightarrow \{1, \ldots, D\}$ *one wants to find an* $x \in \{1, \ldots, M\}$ *such that* $f(x) = \max\limits_{y \in \{1, \ldots, M\}} f(y)$.

There is no better classical algorithm than brute force. The query complexity of the algorithm is $O(M)$. Dürr and Høyer [12] developed a quantum algorithm which is based on the Grover search algorithm. Algorithm 1 describes a variation of their algorithm with expected query complexity $O(\sqrt{M})$. Hereafter we call it $\text{GROVER\_MAX}(f(1), \ldots, f(M))$ and use it as a subroutine for finding an index $x \in \{1, \ldots, M\}$ of the maximal value.

By $\text{GROVER}(h(1), \ldots, h(M))$ we denote a quantum subroutine that equiprobably returns one of $x \in \{1, \ldots, M\}$ such that $h(x) = 1$. If there is no such x, then it returns -1. By $h_\mu : \{1, \ldots, M\} \rightarrow \{0, 1\}$ we denote a function such that $h_\mu(x) = 1$ iff $f(x) > f(\mu)$, for $\mu \in \{1, \ldots, M\}$.

Algorithm 1. $\text{GROVER\_MAX}(f(1), \ldots, f(M))$. Quantum Algorithm for finding an index of the maximal value.

$\mu \leftarrow 1$ ▷ Initially assume 1 as the "best" index
while $\mu \neq -1$ **do**
     $\mu' \leftarrow \mu$ ▷ Remember current "best" index
     $\mu \leftarrow \text{GROVER}(h_{\mu'}(1), \ldots, h_{\mu'}(M))$ ▷ Find a "better" index: $O\left(\frac{M}{\#\{x : f(x) > f(\mu')\}}\right)$
end while
return μ' ▷ There are no elements greater than $f(\mu')$.

Property 1. Algorithm GROVER_MAX by C. Dürr and P. Høyer [12] for maximum search has expected query complexity $O(\sqrt{M})$ and the probability of error at most $\frac{1}{3}$.

3   Balanced Subtraction Games

3.1   Classical Query Complexity

In this subsection we limit our considerations with the number of players $k = 3$, for the sake of simplicity:

* a player who cannot make a move at the end of a play gets 0;
* the previous player who managed to make the last move gets 2;
* the previous-to-previous player gets 1.

All similar results also hold for arbitrary k but require larger formulations, which in our mind are redundant for the understanding.

Lemma 1. *Let Γ be a losing balanced Subtraction game Γ picked uniformly at random. Let game Γ' differ from Γ in exactly one random bit of their binary representations:* HAMMINGDISTANCE $(\Gamma, \Gamma') = 1$.

Then $\Pr\left[\text{WIN}\left(\Gamma'\right) \neq \text{WIN}\left(\Gamma\right)\right] \geq 1/18$.

The idea of the proof of this Lemma resembles the one in [20], but surprisingly $k > 2$ allows to be slightly more simple and efficient. Here, we make three (instead of four) steps when calculating the most meaningful probability (7) and obtain even more strict analogous result. This contrasts with the usual case where generalization to bigger number of players requires more complex analysis.

Proof. First, we mention the following three facts about balanced losing games.

* Balancedness (2) of Γ implies:

$$\Pr_j\left[\text{WIN}\left(\Gamma, j\right) = 1\right] = \Pr_j\left[\text{WIN}\left(\Gamma, j\right) = 0\right] = \frac{1}{3}. \tag{4}$$

* Losingness (1) of Γ implies for each j, $0 < j < n$:

$$\text{either } \Gamma_{nj} = 0 \text{ or } \text{WIN}\left(\Gamma, j\right) = 1, \text{ or both;} \tag{5}$$

otherwise the first player would be able to take $n - j$ stones in the first turn and thus have a positive payoff.

* Assuming Γ to be picked at random, losingness (1) implies

$$\Pr\left[\Gamma_{nj} = 1 \mid \text{WIN}\left(\Gamma, j\right) = 1\right] = \frac{1}{2}, \tag{6}$$

since possibility or impossibility of a (worst possible) move which leads to position "j stones" with $\text{WIN}\left(\Gamma, j\right) = 1$ does not affect the value of any position.

Now let $\Gamma'_{ji} \neq \Gamma_{ji}$ for some pair of indices j, i picked at random, s.t. $0 \leq i < j \leq n$. Then we are interested in the value of

$$\Pr_{j,i}\left[\text{WIN}\left(\Gamma'\right) \neq \text{WIN}\left(\Gamma\right)\right].$$

Let us consider four possible cases for j and i to evaluate this probability. Readers who only care about large values of n, are welcome to skip all but the last case, as the former ones are highly unlikely to happen for random j and i.

1. $j = n$, and $i = 0$.

   Inversion of bit Γ_{n0} changes the value of game Γ with certainty, since Γ is losing (1), so and $\Gamma_{n0} = 0$, but Γ' with $\Gamma'_{n0} = 1$ can be won by taking all n stones in the first turn:

   $$\Pr\left[\text{WIN}\left(\Gamma'\right) \neq \text{WIN}\left(\Gamma\right)\right] = 1$$

2. $j = n$, and $i > 0$.

   (4) $\wedge$ (5) implies that $\Pr\left[\text{WIN}\left(\Gamma, i\right) \neq 1 \wedge \Gamma_{ni} = 0\right] = 1 - \frac{1}{3} = \frac{2}{3}$. Under this condition, inversion of bit Γ_{ni} changes the value of game Γ from 0 to $(\text{WIN}\left(\Gamma, i\right) - 1) \bmod 3 > 0$:

   $$\Pr\left[\text{WIN}\left(\Gamma'\right) \neq \text{WIN}\left(\Gamma\right)\right] = \frac{2}{3}$$

3. $j < n$, and $i = 0$.

   (4) $\wedge$ (5) $\wedge$ (6) implies that $\Pr\left[\text{WIN}\left(\Gamma, j\right) = 1 \wedge \Gamma_{nj} = 1\right] = \frac{1}{3} \times \frac{1}{2} = \frac{1}{6}$ Here $\text{WIN}\left(\Gamma, j\right) = 1$ means that $\Gamma_{j0} = 0$, and inversion of this bit changes the value of $\text{WIN}\left(\Gamma, j\right)$ from 1 to 2. And $\Gamma_{nj} = 1$ means that the value of $\text{WIN}\left(\Gamma\right)$ is at least $(\text{WIN}\left(\Gamma, j\right) - 1) \bmod k$. These two facts together mean that inversion of bit Γ_{j0} changes the value of $\text{WIN}\left(\Gamma, n\right)$ from 0 to 1. Therefore:

   $$\Pr\left[\text{WIN}\left(\Gamma'\right) \neq \text{WIN}\left(\Gamma\right)\right] = \frac{1}{6}$$

4. $0 < i < j < n$.

   (4) $\wedge$ (5) $\wedge$ (6) implies that $\Pr\left[\text{WIN}\left(\Gamma, j\right) = 1 \wedge \text{WIN}\left(\Gamma, i\right) = 0 \wedge \Gamma_{nj} = 1\right] = \frac{1}{3} \times \frac{1}{3} \times \frac{1}{2} = \frac{1}{18}$. Here $\text{WIN}\left(\Gamma, j\right) = 1 \wedge \text{WIN}\left(\Gamma, i\right) = 0$ means that $\Gamma_{ji} = 0$, and inversion of this bit changes the value of $\text{WIN}\left(\Gamma, j\right)$ from 1 to 2. And $\Gamma_{nj} = 1$ means that the value of $\text{WIN}\left(\Gamma\right)$ is at least $(\text{WIN}\left(\Gamma, j\right) - 1) \bmod k$. These two facts together mean that inversion of bit Γ_{ji} changes the value of $\text{WIN}\left(\Gamma, n\right)$ from 0 to 1. Therefore:

   $$\Pr\left[\text{WIN}\left(\Gamma'\right) \neq \text{WIN}\left(\Gamma\right)\right] = \frac{1}{18} \tag{7}$$

In either case we have that $\Pr\left[\text{WIN}\left(\Gamma'\right) \neq \text{WIN}\left(\Gamma\right)\right] \geq \frac{1}{18}$.   □

Theorem 1. *There is no deterministic or randomized algorithm for computing function* WIN *faster than in* $\Omega\left(n^2\right)$ *steps.*

The proof of this theorem is next to the obvious implication of Lemma 1 and the general result is identical, up to the constant values, to the one of [20, Theorem 1]. We refer to the aforementioned work for the details and for the remarks on the eligibility of this proof, which are also relevant here.

3.2   Quantum Algorithm

In this subsection we suggest a quantum algorithm for solving an arbitrary k-player Subtraction game. We assume the reader to be familiar with the basics of quantum computing and, in particular, with the Grover Search algorithm [8,16]. Among other problems, this algorithm is also applicable for searching in directed acyclic graphs (DAGs) [11,19]. In this paper we apply it to Subtraction games which essentially are games on DAGs: if a game-representing binary Γ is treated as an adjacency matrix of DAG $G = (V, E)$, then its set V corresponds to n positions of the game, and its set E corresponds to all the legal moves.

The algorithm determines an optimal move in each possible position, thus providing a strong solution of a game. But as we are interested in the value of $\text{WIN}(\Gamma)$ only, we provide a solution which only returns the value of a game. The algorithm searches for the maximum among directly accessible vertices $\text{ADJ}[j] \overset{\text{def}}{=} \{i : \Gamma_{ji}\}$. This algorithm has two important properties: (i) its expected running time is $O\left(\sqrt{\deg j}\right)$, where $\deg j \overset{\text{def}}{=} |\text{ADJ}[j]|$ is the number of vertices directly accessible from the vertex j; (ii) it returns a vertex i' with the maximal value of WIN with a constant probability (say 0.5) if there exist one or more vertices with maximal values.

In Algorithm 2, we use the Dürr–Høyer Algorithm [12] for minimum search in the form of a GROVER_MAX subroutine which returns maximum value among its arguments. We store the search results in the array w and reuse them in all the subsequent searches.

Algorithm 2. Quantum algorithm for solving a k-player Subtraction game Γ

$w_0 \leftarrow 0$
for $j = 1 \ldots n$ **do** $\triangleright O(n)$
$\quad w_j \leftarrow 0$
$\quad$ **for** $z = 1 \ldots 2 \cdot \lceil \log_2 n \rceil$ **do** $\triangleright O(\log n)$
$\quad\quad w_j \leftarrow \max\left(w_j, \text{GROVER\_MAX}\{(w_i - 1) \bmod k \mid i \in \text{ADJ}[j]\}\right)$ $\triangleright O\left(\sqrt{\deg j}\right)$
$\quad$ **end for**
end for
return w_n

Theorem 2. *Algorithm 2 computes* $\text{WIN}(\Gamma)$ *in expected running time* $O\left(\sqrt{n|E|}\log n\right)$ *and with error probability* $\epsilon \lesssim 1/n$.

Proof. The correctness of the algorithm is obvious: each of the variables w_j (for j running from 1 to n) is assigned a value $\text{WIN}(\Gamma, j)$ according to the definition of the function $\text{WIN}(\Gamma, j)$.

Due to Property 1, the expected running time of GROVER_MAX is $O(\sqrt{\deg j})$ and the error probability is at most $\frac{1}{2}$.

The claim on running time follows from the Cauchy–Bunyakovsky–Schwarz inequality: $\sum_{j=1}^{n} \sqrt{\deg j} \leq \sum_{j=1}^{n} \sqrt{\mathbb{E}_j [\deg j]} = \sum_{j=1}^{n} \sqrt{|E|/n} = \sqrt{n|E|}$.

The probability of error in evaluating one particular w_j is $2^{-2\log_2 n} = 1/n^2$, so the probability of no error at all among evaluations of $w_1, \ldots, w_n$ is $\left(1 - 1/n^2\right)^n \gtrsim 1 - 1/n$. $\qquad\qquad\square$

We note that, for a random Subtraction game, the expected number of edges $\mathbb{E}\big[|E|\big] = \Theta\left(n^2\right)$, and then we conclude that, while the best classical algorithms require time $\Theta\left(n^2\right)$ to solve a Subtraction game, there exists a polynomially faster quantum algorithm which runs in time $O\left(n^{3/2}\log n\right)$.

The exact-time algorithm for a small number of players. If k is a small constant, one can apply a quantum algorithm that works in *exact* time $O(\sqrt{n|E|}\log n)$ (in contrast to Algorithm 2 which has the same evaluation for the *expected* running time). Algorithm 3 runs the Grover Search $k-1$ times instead of running one search for the maximum. At the t-th step the value t is to be searched for among the values from the adjacent vertices, for t running from $(k-1)$ down to 0. Obviously, the first found value is equal to the maximal payoff available in the considered position "j stones". Subroutine GROVER$_t$ in Algorithm 3 searches for the value t among its arguments and returns *True* with probability 0.5 when there is such value, and *False* otherwise. It has to be run $2\cdot\log_2 k\cdot\log_2 n$ times to amplify the probability of success in case if the arguments contain value t.

Algorithm 3. Quantum algorithm for solving a k-player Subtraction game Γ, where k is a small constant

---

$w_0 \leftarrow 0$
for $j = 1\ldots n$ **do**   ▷ $O(n)$
    $w_j \leftarrow 0$
    $t \leftarrow (k-1)$
    while $t > 0 \ \wedge\ w_j = 0$ **do**
        for $z = 1\ldots 2\cdot\lceil\log_2 n\cdot\log_2 k\rceil$ **do**   ▷ $O(\log n)$
            if GROVER$_t\{(w_i - 1) \bmod k \mid i \in \text{ADJ}[j]\}$ **then**   ▷ $O(\sqrt{\deg j})$
                $w_j \leftarrow t$
            end if
        end for
        $t \leftarrow t - 1$
    end while
end for
return w_n

Theorem 3. *If k is a small constant, Algorithm 3 computes* WIN (Γ) *in exact running time* $O\left(\sqrt{n\,|E|}\log n\right)$ *and with error probability* $\epsilon \lesssim 1/n$.

4   Restricted Subtraction Games

4.1   Concept of Solution

Restricted Subtraction games are, in some sense, degenerated, as players essentially have no choice in either position. Nevertheless, these games are of natural interest in terms of the computational complexity of Boolean functions. Namely, they became the first functions formulated in terms of a game, that demonstrate polynomial separation between exact quantum query complexity and classical query complexity.

We note that the adaptive deterministic query complexity of WIN (Γ) for a restricted Subtraction game Γ is n. The upper bound n follows from a very simple analysis of Algorithm 4, the lower bound n also is obvious.

Algorithm 4. Computing WIN (Γ) for a restricted Subtraction game Γ

$w_n \leftarrow 0$
$j \leftarrow n$
for $i = n-1, \ldots, 0$ **do**
    if $\Gamma_{ji} = 1$ **then**
        $j \leftarrow i$
        $w_n \leftarrow (w_n - 1) \bmod k$
    end if
end for
return w_n

Instead of computing WIN (Γ) for a restricted Subtraction game Γ, [18] aims for a more ambitious problem of solving all positions of a game, i.e. of finding vector $W = \left[\text{WIN}(\Gamma, j)\right]_j$.

4.2   Classical Query Complexity

Theorem 4. *Classical query complexity of computing* $\left[\text{WIN}(\Gamma, j)\right]_j$ *for a k-player restricted Subtraction game* Γ *is* $\Theta\left(n^2\right)$.

The proof is identical to one of [18, Theorem 1], which was formulated for the two-player games, but is valid also for the multiplayer games.

4.3   Quantum Algorithm

To solve a restricted Subtraction game, we modify Algorithm 2 according to the idea from [18]. We run the *Exact* Grover Search for a non-zero element in each row. The Exact Grover Search [26] is a modification of Grover's algorithm, which returns the position of the non-zero element in a binary string with Hamming weight 1, or *False* if its Hamming weight is 0. It cannot handle strings with a bigger Hamming weight, but for the promised input, it works in exact time $O\left(\sqrt{n}\right)$ for an n-bit binary string, and with no errors. Subroutine EXACT_GROVER of Algorithm 5 refers to the Exact Grover Search. In contrast to Algorithms 2 and 3, this subroutine has to be called just once as its result is exact and does not need to be amplified.

Algorithm 5. Quantum algorithm for solving a restricted k-player game Γ

$w_0 \leftarrow 0$
for $j = 1 \ldots n$ **do** $\triangleright O(n)$
$\quad t \leftarrow$ EXACT_GROVER $\{$ADJ$[j]\}$ $\triangleright O(\sqrt{\deg j})$
$\quad w_j \leftarrow (w_t - 1) \bmod k$
end for
return w

Theorem 5. *Algorithm 5 computes* WIN(Γ) *in exact running time* $O\left(n^{1.5}\right)$ *and with no error.*

The theorem follows from the properties of the Exact Grover Search and the evaluation of $O(\sqrt{\deg j})$ as in the proof of Theorem 2.

5   Conclusion

Recent results in quantum game theory have stepped into the field of combinatorial games. In this work we generalized several of these results for solving multiplayer combinatorial games. In particular, we established several upper bounds for quantum query complexity, which generally correspond to the running time of a quantum algorithm. We also derived several classical lower bounds for these problems. We did not focus on the classical upper bounds, but they obviously coincide with the lower bounds, which can be shown just by describing the straightforward dynamic programming approach.

The polynomial separation between quantum and classical complexities was shown using different kinds of games and different concepts of solution, but all of them engage Subtraction games for the demonstration of the power of quantum algorithms in combinatorial game theory. Perhaps, one should expect better and more general bounds to emerge for the quantum complexity of solving Subtraction games. Of course, we hope also for detecting other examples of games with quantum-smaller-than-classical complexity.

Acknowledgement. The research is supported by PostDoc Latvia Program, and by the ERDF within the project 1.1.1.2/VIAA/1/16/099 "Optimal quantum-entangled behavior under unknown circumstances". A part of the reported study was funded by RFBR according to the research project No. 19-37-80008. A part of the research was funded by the subsidy allocated to Kazan Federal University for the state assignment in the sphere of scientific activities, project No. 0671-2020-0065.

References

1. Ablayev, F., Ablayev, M., Zhexue, H.J., Khadiev, K., Salikhova, N., Wu, D.: On quantum methods for machine learning problems part I: quantum tools. Big Data Min. Anal. **3**(1), 41–55 (2019)

2. Ambainis, A.: Understanding quantum algorithms via query complexity. arXiv preprint arXiv:1712.06349 (2017)
3. Ambainis, A., et al.: Quantum strategies are better than classical in almost any XOR game. In: Czumaj, A., Mehlhorn, K., Pitts, A., Wattenhofer, R. (eds.) ICALP 2012. LNCS, vol. 7391, pp. 25–37. Springer, Heidelberg (2012). https://doi.org/10.1007/978-3-642-31594-7_3
4. Ambainis, A., Iraids, J., Kravchenko, D., Virza, M.: Advantage of quantum strategies in random symmetric XOR games. In: Kučera, A., Henzinger, T.A., Nešetřil, J., Vojnar, T., Antoš, D. (eds.) MEMICS 2012. LNCS, vol. 7721, pp. 57–68. Springer, Heidelberg (2013). https://doi.org/10.1007/978-3-642-36046-6_7
5. Ardehali, M.: Bell inequalities with a magnitude of violation that grows exponentially with the number of particles. Phys. Rev. A **46**, 5375–5378 (1992)
6. Benjamin, S.C., Hayden, P.M.: Comment on "quantum games and quantum strategies". Phys. Rev. Lett. **87**(6), 069801 (2001)
7. Benjamin, S.C., Hayden, P.M.: Multi-player quantum games. Phys. Rev. A **64**(3), 030301 (2001)
8. Boyer, M., Brassard, G., Høyer, P., Tapp, A.: Tight bounds on quantum searching. Fortschritte der Physik **46**(4–5), 493–505 (1998)
9. Brunner, N., Linden, N.: Connection between Bell nonlocality and Bayesian game theory. Nat. Commun. **4**, 2057 (2013)
10. Clauser, J.F., Horne, M.A., Shimony, A., Holt, R.A.: Proposed experiment to test local hidden-variable theories. Phys. Rev. Lett. **23**, 880 (1969)
11. Cormen, T.H., Leiserson, C.E., Rivest, R.L., Stein, C.: Introduction to Algorithms, 2nd edn. McGraw-Hill, New York (2001)
12. Dürr, C., Høyer, P.: A quantum algorithm for finding the minimum. arXiv:quant-ph/9607014 (1996)
13. Eisert, J., Wilkens, M., Lewenstein, M.: Quantum games and quantum strategies. Phys. Rev. Lett. **83**, 3077 (1999)
14. van Enk, S.J., Pike, R.: Classical rules in quantum games. Phys. Rev. A **66**, 024306 (2002)
15. Ferguson, T.S.: Game theory class notes for math 167, fall 2000 (2000). https://www.cs.cmu.edu/afs/cs/academic/class/15859-f01/www/notes/comb.pdf
16. Groverm, L.K.: A fast quantum mechanical algorithm for database search. In: Proceedings of the Twenty-eighth Annual ACM Symposium on Theory of Computing, pp. 212–219. ACM (1996)
17. Grundy, P.M.: Mathematics and games. Eureka **2**, 6–8 (1939)
18. Huang, Y., Ye, Z., Zheng, S., Li, L.: An exact quantum algorithm for a restricted subtraction game. Int. J. Theoret. Phys. **59**(5), 1504–1511 (2020). https://doi.org/10.1007/s10773-020-04418-z
19. Khadiev, K., Safina, L.: Quantum algorithm for dynamic programming approach for DAGs. Applications for Zhegalkin polynomial evaluation and some problems on DAGs. In: McQuillan, I., Seki, S. (eds.) UCNC 2019. LNCS, vol. 11493, pp. 150–163. Springer, Cham (2019). https://doi.org/10.1007/978-3-030-19311-9_13
20. Kravchenko, D., Khadiev, K., Serov, D.: On the quantum and classical complexity of solving subtraction games. In: van Bevern, R., Kucherov, G. (eds.) CSR 2019. LNCS, vol. 11532, pp. 228–236. Springer, Cham (2019). https://doi.org/10.1007/978-3-030-19955-5_20
21. Khan, F.S., Humble, T.S.: Nash embedding and equilibrium in pure quantum states. In: Feld, S., Linnhoff-Popien, C. (eds.) QTOP 2019. LNCS, vol. 11413, pp. 51–62. Springer, Cham (2019). https://doi.org/10.1007/978-3-030-14082-3_5

22. Khan, F.S., Solmeyer, N., Balu, R., Humble, T.S.: Quantum games: a review of the history, current state, and interpretation. Quantum Inf. Process. **17**(11), 1–42 (2018). https://doi.org/10.1007/s11128-018-2082-8
23. Kravchenko, D.: A new quantization scheme for classical games. In: Proceedings of Workshop on Quantum and Classical Complexity (Satellite event to ICALP 2013), pp. 17–34 (2013)
24. Kravchenko, D.: Quantum entanglement in a zero-sum game. Contrib. Game Theory Manag. **8**, 149–163 (2015)
25. Li, Y.D.: BQP and PPAD. arXiv:1108.0223 (2011)
26. Long, G.-L.: Grover algorithm with zero theoretical failure rate. Phys. Rev. A **64**(2), 022307 (2001)
27. Marinatto, L., Weber, T.: A quantum approach to static games of complete information. Phys. Lett. A **272**, 291–303 (2000)
28. Mermin, D.: Extreme quantum entanglement in a superposition of macroscopically distinct states. Phys. Rev. Lett. **65**, 15 (1990)
29. Meyer, D.A.: Quantum strategies. Phys. Rev. Lett. **82**, 1052–1055 (1999)
30. Muhammad, S., Tavakoli, A., Kurant, M., Pawlowski, M., Zukowski, M., Bourennane, M.: Quantum bidding in bridge. Phys. Rev. X **4**, 021047 (2014)
31. Nielsen, M.A., Chuang, I.L.: Quantum Computation and Quantum Information. Cambridge University Press, Cambridge (2010)
32. Phoenix, S.J.D., Khan, F.S.: Preferences in quantum games. Phys. Lett. A **384**(15) (2020)
33. Phoenix, S.J.D., Khan, F.S.: The role of correlation in quantum and classical games. Fluct. Noise Lett. **12**(3), 1350011 (2013)
34. Roch, C., et al.: A quantum annealing algorithm for finding pure Nash equilibria in graphical games. In: Krzhizhanovskaya, V.V., et al. (eds.) ICCS 2020. LNCS, vol. 12142, pp. 488–501. Springer, Cham (2020). https://doi.org/10.1007/978-3-030-50433-5_38
35. Sprague, R.P.: Über mathematische kampfspiele. Tohoku Math. J. **41**, 438–444 (1935)
36. Vaidman, L.: Variations on the theme of the Greenberger-Horne-Zeilinger proof. Found. Phys. **29**, 615 (1999). https://doi.org/10.1023/A:1018868326838
37. Werner, R.F., Wolf, M.M.: Bell inequalities and entanglement. Quantum Inf. Comput. **1**(3), 1–25 (2001)
38. Werner, R.F., Wolf, M.M.: All multipartite Bell correlation inequalities for two dichotomic observables per site. Phys. Rev. A **64**, 032112 (2001)
39. Zhang, S.: Quantum Strategic Game Theory. arXiv:1012.5141 (2010)

Efficient Restrictions of Immediate Observation Petri Nets

Michael Raskin$^{(\boxtimes)}$◉ and Chana Weil-Kennedy◉

Technical University of Munich, Munich, Germany
{raskin,chana.weilkennedy}@in.tum.de

Abstract. In a previous paper we introduced immediate observation Petri nets [9], a subclass of Petri nets with application domains in distributed protocols and theoretical chemistry (chemical reaction networks). IO nets enjoy many useful properties [9,14], but like the general case of conservative Petri nets they have a PSPACE-complete reachability problem. In this paper we explore two restrictions of the reachability problem for IO nets which lower the complexity of the problem drastically. The complexity is NP-complete for the first restriction with applications in distributed protocols, and it is polynomial for the second restriction with applications in chemical settings.

Keywords: Petri nets · Reachability · Computational complexity

1 Introduction

In this paper we refine our results about the complexity of verifying immediate observation Petri nets [9] in the case of two restrictions of such nets. Petri nets and their subclasses are widely used and studied in the context of software and system verification (e.g. [7]), but also others such as game theory (e.g. [11]), chemical reaction networks (e.g. [3]) etc. Unfortunately many important problems there have high complexity, and reachability is at least TOWER-hard in the general case [6]. This motivates the study of subclasses of Petri nets.

Immediate observation Petri nets (IO nets) are a reformulation of immediate observation population protocols, which have been introduced by Angluin et al. in [2]. Initially, they were studied from the point of view of computing predicates in a distributed system, where their expressive power is lower than general population protocols (conservative Petri nets) but still considerable. Many verification problems for IO nets are PSPACE-complete; among them set-parametrized problems for sets defined by boolean combinations of bounds on token counts. This is a significant improvement compared to the general or conservative case of Petri nets, where EXPSPACE-hard [4] and even harder verification problems are the

This project has received funding from the European Research Council (ERC) under the European Union's Horizon 2020 research and innovation programme under grant agreement No 787367 (PaVeS).

© Springer Nature Switzerland AG 2020
S. Schmitz and I. Potapov (Eds.): RP 2020, LNCS 12448, pp. 99–114, 2020.
https://doi.org/10.1007/978-3-030-61739-4_7

norm. IO nets provide a natural description of some distributed systems, but also can be used to describe enzymatic chemical networks [1].

Of course, a subclass of reachability problems with a better computational complexity raises some natural, even if informal, questions. What allows better complexity and can it be generalized to some wider subclass? What keeps the complexity from being even lower and are there useful subclasses without these obstacles? Are there applications where a typical problem can be solved more efficiently? We believe that branching immediate observation nets, a generalization of IO nets and basic parallel processes with reachability problem in PSPACE [14], answer the first question. The present paper is devoted to the last two questions.

We consider two restrictions, the first one a syntactic restriction defining a subclass of IO nets, and the second a condition on the initial and final markings considered in the reachability problem for IO nets. The first restriction is plausible in some distributed systems, and it also bears similarity to the delayed observation population protocols introduced by Angluin et al. in [2]. The second restriction has applications in some chemical systems (enzymatic chemical reaction networks, [1]). We show the first restriction entails an NP-complete reachability problem, and for the second restriction we provide a polynomial algorithm deciding reachability or giving a witness that the restriction does not hold.

The rest of the paper is organized as follows. In Sect. 2, we recall some general definitions regarding Petri nets, as well as the classic maximum flow minimum cut problem. Section 3 defines immediate observation Petri nets. Then we show the effects for reachability complexity of two restrictions on IO nets: keeping transitions enabled once enabled in Sect. 4, and requiring all token counts and their combinations to be large or zero in Sect. 5. Finally, we summarize our results in the conclusion and outline some further directions.

Due to space restrictions some of the technical details of the proofs have been omitted and can be found in the full version [13].

2   Preliminaries

Multisets. A *multiset* on a finite set E is a mapping $C\colon E \to \mathbb{N}$, i.e. for any $e \in E$, $C(e)$ denotes the number of occurrences of element e in C. Let $\langle e_1, \ldots, e_n \rangle$ denote the multiset C such that $C(e) = |\{j \mid e_j = e\}|$. Operations on $\mathbb{N}$ like addition or comparison are extended to multisets by defining them component wise on each element of E. Given $X \subseteq E$ define $C(X) \stackrel{\text{def}}{=} \sum_{e \in X} C(e)$. We call $\sum_{e \in E} C(e)$ the *size* of C and note it $|C|$.

Place/Transition Petri Nets with Weighted Arcs. A *Petri net* N is a triple (P, T, W) consisting of a finite set of *places* P, a finite set of *transitions* T and a *weight function* $W\colon (P \times T) \cup (T \times P) \to \mathbb{N}$. A *marking* M is a multiset on P, and we say that a marking M puts $M(p)$ *tokens* in place p of P. The *size* of M, denoted by $|M|$, is the total number of tokens in M. The *preset* ${}^{\bullet}t$ and *postset* $t^\bullet$ of a transition t of T are the multisets on P given by $^\bullet t(p) = W(p,t)$ and $t^\bullet(p) = W(t,p)$. A transition t is *enabled* at a marking M if $^\bullet t \leq M$, i.e. $^\bullet t$ is component-wise smaller or equal to M. If t is enabled then it can be *fired*, leading to a new marking $M' = M - {}^\bullet t + t^\bullet$. We let $M \xrightarrow{t} M'$ denote this. Given $\sigma = t_1 \ldots t_n$ we write $M \xrightarrow{\sigma} M_n$ when $M \xrightarrow{t_1} M_1 \xrightarrow{t_2} M_2 \ldots \xrightarrow{t_n} M_n$, and call σ a *firing sequence*. We write $M' \xrightarrow{*} M''$ if $M' \xrightarrow{\sigma} M''$ for some $\sigma \in T^*$, and say that M'' is *reachable* from M'.

Flows and Cuts. A *flow graph* is a triple $G = (V, A, c)$ where V is a finite set of vertices, $A \subseteq V^2$ is a finite set of arcs, and $c : A \to \mathbb{N} \cup \{\infty\}$ is a nonnegative *capacity* function on arcs. Given an arc $a \in A$, we call $c(a)$ the *capacity* of a. Notice that this capacity can be infinite. A flow graph contains two special vertices i and o, called the *inlet* and *outlet*, such that i has no incoming arc and o has no outgoing arc. A *flow* of a flow graph is a function $f : A \to \mathbb{N}$ such that $f(a) \leq c(a)$ for each arc $a \in A$, and for each vertex $v \in V \setminus \{i, o\}$, the sum of the flow over v's incoming arcs is equal to the sum of the flow over v's outgoing arcs. The *value* of a flow is the sum $\sum_{(i,p) \in A} f((i,p))$ of the flow over all arcs from the inlet, or equivalently the sum $\sum_{(p,o) \in A} f((p,o))$ of the flow over all arcs to the outlet. A *cut* in a flow graph $G = (V, A, c)$ is a pair of disjoint subsets $V_I \sqcup V_O = V$ such that the inlet is in V_I and the outlet is in V_O. The *capacity* of a cut (V_I, V_O) is the sum of the capacities of all the arcs going from vertices in V_I to vertices in V_O. We say an arc $a = (u, v)$ *crosses the cut*, if $u \in V_i$ and $v \in V_O$.

We recall two classic theorems.

Theorem 1 (Max-flow min-cut theorem [10]). *In a flow graph, the maximum value of a flow is equal to the minimum capacity of a cut.*

Theorem 2 (Dinitz algorithm [8]). *Given a flow graph, a flow with the maximum value and a cut with the minimum capacity can be found in polynomial time.*

3 Immediate Observation Petri Nets

We recall the definition of immediate observation nets (IO nets) from [9].

Definition 1. *A transition t of a Petri net is an* immediate observation transition *(IO transition) if there are places p_s, p_d, p_o, not necessarily distinct, such that $^\bullet t = \{p_s, p_o\}$ and $t^\bullet = \{p_d, p_o\}$. We call p_s, p_d, p_o the* source, destination, *and* observed *places of t, respectively. We denote by $p_s \xrightarrow{p_o} p_d$ such a transition. A Petri net is an* immediate observation net *(IO net) if all its transitions are IO transitions.*

Following the graphical convention of [12] for contextual nets, we represent the Petri net arcs (p_o, t) and (t, p_o) by an undirected arc between t and p_o in our figures. This emphasizes that transition t has a read-only relation to its observed place p_o. In the examples, we also consider IO nets containing transitions with no observed place. To make the net a formally correct IO net, it suffices to add an extra marked place which acts as observed place for these transitions.

IO nets are *conservative*, i.e. there is no creation or destruction of tokens.

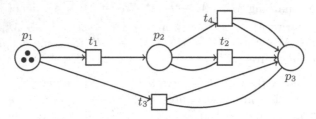

Fig. 1. An IO net.

Example 1. Figure 1 shows an IO net taken from the literature on population protocols [2]. Intuitively, it models a protocol allowing a crowd of undistinguishable agents that can only interact in pairs to decide whether they are at least 3. Given a marking M_0 with tokens only in p_1, if $M_0(p_1) \geq 3$, then repeated firing of an arbitrary enabled transition eventually puts all the tokens into p_3.

In [9], we showed that given an IO net N and two markings M, M', deciding whether M' is reachable from M is a PSPACE-complete problem. The proof of PSPACE-hardness for the reachability problem in IO nets uses a reduction from the halting problem of linear-space Turing machines. The reduction is done by simulating the runs of the Turing machine: places describe the state of the head and of the tape cells, and transitions model the movement of the head and the change in the symbols on the tape cells. In the construction a specific "success" place becomes marked if and only if the machine reaches the halting state without exceeding the permitted space.

The nets provided by this reduction have two common properties. First, the transitions get enabled and disabled a large number of times. Second, the markings put at most one token per place. We show how forcing a strong enough contrary condition to at least one of these properties leads to much easier verification.

4   First Restriction: Transition Enabling

The PSPACE-hardness proof for IO reachability relies on the observation requirements of some transitions switching between satisfied and unsatisfied many times. In some distributed systems, observations correspond to irrevocable declarations of the agents, for example in some multi-phase commit protocols. We consider IO nets where a token move enabled by observing some token remains enabled even when the observed token has changed places. We formalize such a property in the following definition.

Definition 2. *An IO net is* non-forgetting *if for each transitions $p \xrightarrow{r} q$ and $r \xrightarrow{s} r'$ there is also a transition $p \xrightarrow{r'} q$.*

Consider a marking of an IO net where the observation place of some transition with source place p and destination place q is marked. If there is a token in place p, then it can move to q. We say that the *token move from p to q is enabled*. In a non-forgetting IO net, once the token move from p to q is enabled in some marking of a firing sequence, it stays enabled in the subsequent markings of the firing sequence. Notice that the token move from p to q being enabled in a marking is not equivalent to a transition from p to q being enabled: a transition is enabled when both its observation place and its source place are marked, whereas a token move is enabled as soon as the observation place of some suitable transition is marked.

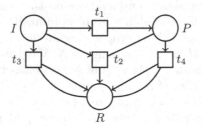

Fig. 2. A non-forgetting Petri net.

Example 2. The non-forgetting IO net of Fig. 2 models one of the steps of updating a shared state: A proposal can be published and stored, and every agent has an opportunity to veto it.

All agents start in the initial state I. Some agent can propose a change by moving from state I to state P. If there is a proposal, an agent can move from state I to state P to support the proposal, or go to the state R to reject the proposal. If there is an agent rejecting the proposal (i.e. in the state R), other agents can move to R both from I and from P to recognise the fact that the proposal has been rejected. Note that the agents cannot reject a proposal before it has been created, which is encoded by P being the observed place of t_2. Also note that the agent proposing a change cannot start rejecting it until some other agent rejects it.

The reachability problem for such IO nets becomes much simpler.

Theorem 3. *The reachability problem for non-forgetting IO nets is in* NP.

Proof. Let N be a non-forgetting IO net. Consider a (non-empty) firing sequence σ of N from markings M to M'. It can be decomposed into n non-empty subsequences σ_i such that $M = M_0 \xrightarrow{\sigma_1} M_1 \xrightarrow{\sigma_2} M_2 \ldots \xrightarrow{\sigma_n} M_n = M'$ for some $n > 0$, and such that M_i are the markings of the firing sequence in which new token moves become enabled. Recall that since N is non-forgetting, a token move once enabled remains enabled. There are at most $|P|^2$ such subsequences in any firing sequence, and in each subsequence the set of enabled token moves is fixed.

Example 3. Consider the net of Example 2, and the firing sequence $t_2^3 t_4$ from marking $(4, 1, 0)$, which put 4 tokens in I, 1 tokens in P and 0 token in R, to marking $(1, 0, 4)$. This firing sequence is decomposed into two subsequences: $(4, 1, 0) \xrightarrow{t_2} (3, 1, 1)$ and $(3, 1, 1) \xrightarrow{t_2^2 t_4} (1, 0, 4)$. In the first, the token moves from I to R and from I to P are enabled. In the second, these token moves as well as the token move from P to R are enabled.

To show that the reachability problem for non-forgetting IO nets is in NP, we define a reachability certificate and show how it can be verified in polynomial time. The certificate corresponding to a firing sequence consists of the markings in which some token move is enabled for the first time. Such a certificate has polynomial length by the above considerations on the number of subsequences.

We now show that the reachability problem in an IO net with a fixed set of enabled token moves is reducible to the maximum flow problem on graphs. Let N be an IO net, let M, M' be two markings of N. We define G as the flow graph with vertices identified with the places P of N, as well as two additional vertices i and o, the inlet and outlet of the flow graph.

For each enabled token move from p to q for some places p, q, there is an arc from p to q in G with infinite capacity. Each vertex p identified with a place of N has one incoming arc from the inlet i with capacity $M(p)$, and one outgoing arc to the outlet o with capacity $M'(p)$.

Example 4. Figure 3 illustrates two such flow graphs for the non-forgetting IO net of Example 2. The first flow graph corresponds to the enabled token moves from I to R and from I to P, with markings $M = (4, 1, 0)$ and $M' = (3, 1, 1)$. The second flow graph corresponds to the enabled token moves from I to R, from I to P and from P to R, with markings $M = (3, 1, 1)$ and $M' = (1, 0, 4)$.

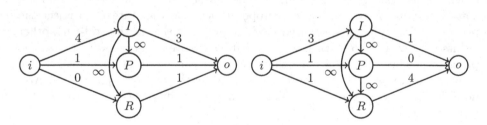

Fig. 3. Flow graphs corresponding to the non-forgetting net of Fig. 2.

A firing sequence σ from M to M' in N corresponds naturally to an integer flow f on G, where for all vertices p and q corresponding to places of the IO net, $f(i, p) = M(p), f(p, o) = M'(p)$ and $f(p, q)$ is equal to the number of transitions from p to q in σ. This flow has value $|M| = |M'|$.

Conversely, an integer flow of value $|M| = |M'|$ corresponds to a firing sequence in N, provided N has a fixed set of enabled token moves. Let us consider such a flow f. It corresponds to a multiset θ of token moves. Starting with the marking M, we remove from the multiset some token move with the source place having more tokens than in M' and fire some corresponding enabled transition. We continue until we reach M'. The details of the construction and its correctness proof are purely technical and can be found in the full version.

We see that verifying a certificate requires a polynomial number of invocations of a polynomial-time algorithm. This concludes the proof.

In fact the reachability problem is NP-complete.

Theorem 4. *Reachability problem for non-forgetting IO nets is* NP-*hard.*

Proof (Sketch). NP-hardness of reachability is proved by a reduction from the NP-complete SAT problem. Consider a SAT instance represented as a circuit of binary "NAND" $(\neg(x \wedge y))$ operations. One can construct a net such that its runs correspond to the input nodes of the circuit choosing arbitrary input values, and the operation nodes of the circuit evaluating the function given the chosen values of the inputs. The technical details are provided in the full version.

5   Second Restriction: Token Counts

Another property of the PSPACE-hardness reduction for IO nets is the low number of tokens in each place. Specifically, no reachable marking puts more than one token in any place. Some systems exhibit a very different behaviour. For instance in most cases of chemical reaction networks, the number of individual molecules is much larger than the number of species of molecules. Additionally, we do not expect any chance "near-misses" between the configuration of the molecules before and after a reaction sequence. If the total amount of molecules of some group of species before the reaction sequence is approximately equal to the amount of molecules of some other group of species afterwards, there must be a precise equality following from some conservation laws.

This behaviour can be formalized by the following condition.

Definition 3. *A pair of markings M and M' of an IO net of place set P is a* near-miss *pair if there exists sets of places X and Y such that $0 < |M(X) - M'(Y)| \leq |P|^3$. A pair which is not a near-miss is called a* no-near-miss *pair.*

Observe that each place of markings M and M' such that M, M' are a no-near-miss pair can be either unmarked or contain at least $|P|^3$ tokens. This can be seen by examining sets $X = \{p\}$ and $Y = \emptyset$, or $X = \emptyset$ and $Y = \{p\}$ in the definition.

Example 5. Consider the IO net of Fig. 4 which models a system where an enzyme E can be produced by an enzyme producer PE, and where a resource molecule R can transform into a product molecule $P1$ in the presence of an enzyme E, or into a product molecule $P2$. On the one hand, the total amount of the two products $P1$ and $P2$ together must match the amount of resource R consumed; on the other hand, it would be surprising if the two products were produced in the same amounts with high but imperfect precision, as there is

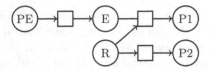

Fig. 4. An example of an IO net with enzyme production and use.

nothing ensuring such an approximate equality. Informally, we can consider the scales from an example of [5] cited in [1]. Five species of molecules are considered in a milliliter-scale cell (although with a different net which is not immediate observation). The concentrations of molecules are measured in picomoles per milliliter. As a picomole contains more than 10^{11} molecules, equalities that hold up to 10^3 molecules have a relative error of 10^{-8}. Such equalities might be expected to follow from some conservation laws and be precise.

Theorem 5. *The IO net reachability problem for no-near-miss pairs of markings is in P. Moreover, there is a polynomial-time algorithm such that for every pair of markings M, M' it either resolves reachability, giving a witness firing sequence if it exists, or reports a near-miss in M and M'.*

Even though the no-near-miss property is NP-complete (e.g. via SUBSET-SUM), making a proof of its violation an alternative valid answer of the algorithm simplifies IO reachability.

Remark 1. Requiring only that the initial and final markings of a firing sequence have many tokens in the non-empty places does not give us a better complexity than the general PSPACE-complete case.

Example 6. Consider two markings on the net of Fig. 4, M with 200 tokens in PE and 400 tokens in R, and M' with 200 tokens in E and 400 tokens in $P1$. The pair (M, M') is a no-near-miss, and we will illustrate the algorithm by verifying reachability from M to M'.

The core idea of the algorithm is to maintain an increasing set of restrictions. Once there are no restrictions to add, we either construct a firing sequence from M to M' satisfying the obtained restrictions and no other ones, use the restrictions to prove that M cannot reach M', or find a near-miss in M and M'.

5.1 Restrictions

We first recall some definitions from [9], and then describe our restrictions and what it means for a restriction set to be stable.

Trajectories and Histories. Since the transitions of IO nets do not create or destroy tokens, we can give tokens identities. Given a firing sequence, each token of the initial marking follows a *trajectory* through the places of the net until it reaches the final marking of the sequence. The trajectories of the tokens between given source and target markings constitute a *history*.

A *trajectory* of IO net N is a sequence $\tau = p_1 \dots p_k$ of places. We let $\tau(i)$ denote the i-th place of τ. The *i-th step* of τ is the pair $\tau(i)\tau(i+1)$. A *history H* of length h is a multiset of trajectories of length h. Given an index $1 \leq i \leq h$, *the i-th marking of H*, denoted M_H^i, is defined as follows: for every place p, $M_H^i(p)$ is the number of trajectories $\tau \in H$ such that $\tau(i) = p$. The markings M_H^1 and M_H^h are the *initial* and *final* markings of H, and we write $M_H^1 \xrightarrow{H} M_H^h$. A history H of length $h \geq 1$ is *realizable* if there exist transitions $t_1, \dots, t_{h-1}$ and numbers $k_1, \dots, k_{h-1} \geq 0$ such that

- $M_H^1 \xrightarrow{t_1^{k_1}} M_H^2 \cdots M_H^{h-1} \xrightarrow{t_{h-1}^{k_{h-1}}} M_H^h$, where for every t we define $M' \xrightarrow{t^0} M$ iff $M' = M$.
- For every $1 \leq i \leq h - 1$, there are exactly k_i trajectories $\tau \in H$ such that $\tau(i)\tau(i+1) = p_s p_d$, where p_s, p_d are the source and target places of t_i, and all other trajectories $\tau \in H$ satisfy $\tau(i) = \tau(i+1)$. Moreover, there is at least one trajectory τ in H such that $\tau(i)\tau(i+1) = p_o p_o$, where p_o is the observed place of t_i. We say that t_i *realizes* step i of H.

We say that $t_1^{k_1} \cdots t_{h-1}^{k_{h-1}}$ realizes H. Intuitively, at a step of a realizable history only one transition occurs, although perhaps multiple times, for different tokens. From the definition of realizable history we immediately obtain:

- $M' \xrightarrow{*} M$ iff there exists a realizable history with M' and M as initial and final markings.
- Every firing sequence that realizes a history of length h has accelerated length at most h.

Restriction Definition. Given an IO net N, places p, q, r of N, and two markings M and M', we say that *a token goes from p to q via r* if there exists a realizable history H of length h between M and M' and a trajectory τ in H such that $\tau(1) = p$, $\tau(h) = q$ and $\tau(i) = r$ for some $i \in \{1, \dots, h\}$.

Given a pair M, M', our algorithm computes a set $\mathcal{R}$ of *restrictions* of the form (p, r, q). We say a restriction (p, r, q) is *correct* if no token goes from p to q via r, i.e. if there is no realizable history from M to M' containing a trajectory from p to q passing through r. We say that a pair of places (p, q) is *forbidden* if for all $r \in P$ the restriction (p, r, q) is in $\mathcal{R}$. *Forbidding* a pair (p, q) means adding the restriction (p, r, q) to $\mathcal{R}$ for all $r \in P$. A pair of places (p, q) that is not forbidden is *allowed*.

Flow Graph. We define a correspondence between the reachability problem in an IO net with a (correct) restriction set and the maximum flow problem for a certain flow graph.

Let N be an IO net of place set P, let M, M' be two markings of N, and let $\mathcal{R}$ be a set of restrictions. We define the *flow graph* $G = (V, A, c)$ with $2|P| + 2$ vertices. There are two vertices for each place $p \in P$, an "initial" copy v_p^i and a "final" copy v_p^f, as well as a distinguished inlet vertex i and a distinguished outlet vertex o. For each place $p \in P$, there is an arc $a = (i, v_p^i)$ with capacity $c(a) = M(p)$, and an arc $a = (v_p^f, o)$ with capacity $c(a) = M'(p)$. For each pair of places $(p, q) \in P^2$ such that (p, q) is allowed in $\mathcal{R}$, there is an arc $a = (v_p^i, v_q^f)$ from the initial p-labeled vertex to the final q-labeled vertex with infinite capacity. Note that the maximum flow value in graph G thus constructed is at most $|M| = |M'|$.

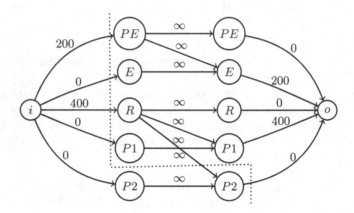

Fig. 5. Flow graph for the IO net of Fig. 4 with a cut.

Example 7. Figure 5 illustrates the flow graph G constructed for the IO net of Fig. 4, the markings $M = (200, 0, 400, 0, 0)$ and $M' = (0, 200, 0, 400, 0)$, and the restriction set that allows only pairs of the form (p, p) and also the pairs (PE, E), $(R, P1)$, $(R, P2)$.

A realizable history H from M to M' naturally corresponds to a flow of value $|M|$: the flow that saturates all the arcs with the finite capacities (i.e. the arcs from the inlet and to the outlet), and assigns to an infinite-capacity arc from v_p^i to v_q^f the number of trajectories from p to q in H. Since this flow saturates all the finite edges, it is a maximum flow.

Stable Restriction Set. We define the notion of a stable set of restrictions for a pair of marking M and M'. Intuitively, a stable set of restrictions does not immediately exclude reachability from M to M', and cannot be extended.

Definition 4. *A set $\mathcal{R}$ of correct restrictions for an IO net N and configurations M and M' is stable if the following conditions hold.*

1. *The maximum flow in the corresponding flow graph is equal to the size $|M|$ of the configurations M (and M').*
2. *For each two places p and q, if there is a minimum cut of the flow graph with v_p^i in the outlet component and v_q^f in the inlet component, the pair (p, q) is forbidden.*
3. *For each larger set of restrictions $\mathcal{R}' \supsetneq \mathcal{R}$, either there is a pair (p, q) such that the triple $(p, p, q) \in \mathcal{R}' \setminus \mathcal{R}$, or there is exists a transition $s \xrightarrow{o} d$ and triples $(p, s, q), (p', o, q') \notin \mathcal{R}'$ and $(p, d, q) \in \mathcal{R}' \setminus \mathcal{R}$.*
4. *For each larger set of restrictions $\mathcal{R}' \supsetneq \mathcal{R}$, either there is a pair (p, q) such that the triple $(p, q, q) \in \mathcal{R}' \setminus \mathcal{R}$, or there exist a transition $s \xrightarrow{o} d$ and triples $(p, d, q), (p', o, q') \notin \mathcal{R}'$ and $(p, s, q) \in \mathcal{R}' \setminus \mathcal{R}$.*

Each of these conditions prohibits some property that can rule out reachability or imply new restrictions. We give some intuition now, then prove formally in Sect. 5.2 that in the case where M and M' are a no-near-miss pair, we can build a realizable history from M to M' from a stable set of restrictions. Moreover the history constructed will show that the set of restrictions cannot be extended.

We call the first two conditions *flow-based stability conditions*. The first condition corresponds to the fact that if a restriction set leads to a flow graph with a maximum flow smaller than $|M|$, then there can be no realizable history from M to M' consistent with such restrictions. The second condition uses the fact that a minimum cut has the same value as a maximum flow, which has size $|M|$ by the first flow-based condition. Let (p, q) be a pair violating the condition. A max flow f that uses the edge from v_p^i to v_q^f can be decomposed into a sum of two flows f_1 and f_2: f_1 the flow with value 1 along path $i - v_p^i - v_q^f - o$ and $f_2 = f - f_1$ which has value $|M| - 1$. Flow f_1 uses two arcs of the minimum cut thus yielding a contradiction by leaving a cut of capacity $|M| - 2$ to f_2. This contradicts existence of a maximum flow using the edge from v_p^i to v_q^f and thus the existence of a realizable history from M to M' with trajectories from p to q.

Example 8. Figure 5 illustrates a minimal cut on the flow graph G of Example 7 in which the path $i \to v_R^i \to v_{P2}^f \to o$ contains two arcs crossing the cut. The restriction set is not stable and $(R, P2)$ must be forbidden.

We call the last two conditions *reachability-based stability conditions*. They rule out an inductive proof of a larger restriction set in the following sense. Given a larger set $\mathcal{R}'$ which violates one of these conditions, we will show by induction on the step number that any realizable history deduced from $\mathcal{R}$ is also coherent with $\mathcal{R}'$, and thus we can replace $\mathcal{R}$ with the larger set $\mathcal{R}'$.

5.2 Firing Sequence Construction

We show how to construct a firing sequence from a stable restriction set, possibly reporting a near-miss instead. The proof that the near-miss reports are correct is after the construction, in Sect. 5.3.

Given a flow graph $G = (V, A, c)$, we define two operations on the capacity c relative to a place pair $(p, q) \in P^2$ and an integer $k > 0$. *Increasing c by k along*

(p, q) consists in increasing $c(i, v_p^i)$ and $c(v_q^f, o)$ by k. *Decreasing c by k along* (p, q) consists in decreasing $c(i, v_p^i)$ and $c(v_q^f, o)$ by k. This decreasing operation is not possible if $c(i, v_p^i)$ or $c(v_q^f, o)$ are smaller than k.

From Stable Restriction Set to Solution Flow. Given a stable set of restrictions with b allowed pairs (p, q), a *solution flow* is a result of the following procedure: Construct the flow graph G. Decrease the capacity by $|P|$ along each allowed pair; if this step fails because some arc has insufficient capacity, terminate the algorithm and report that M, M' is a near-miss pair. Otherwise, compute a maximal flow. If it has value less than $|M| - b \times |P|$, terminate the algorithm and report that M, M' is a near-miss pair. Otherwise, increase its capacity by $|P|$ along each (allowed) pair.

Example 9. In our running example, consider a stable set of restrictions $\mathcal{R}$ allowing only the triples (PE, PE, E), (PE, E, E), $(R, R, P1)$, and $(R, P1, P1)$. This corresponds to a solution flow assigning the edges of the path $i \to PE \to E \to o$ the value 200 and the edges of the path $i \to R \to P1 \to o$ the value 400.

Observe that when a solution flow exists, it might not be unique. The algorithm builds a firing sequence from the solution flow.

From Solution Flow to Firing Sequence. Let $\mathcal{R}$ be a stable restriction set of the algorithm, and let f be a corresponding solution flow. Intuitively, our construction of the solution flow makes sure the flow has value at least $|P|$ along each pair (p, q) allowed by $\mathcal{R}$. We use the reachability-based stability conditions to construct a realizable history from this flow, such that for every pair (p, q) there are at most $f(v_p^i, v_q^f)$ trajectories from p to q.

We define three markings M_m, M_i and M_f. We denote $A(p, q)$ the set of all places r such that the triple (p, r, q) is allowed, i.e. $(p, r, q) \notin \mathcal{R}$. Let M_m be the marking such that $M_m(r)$ is equal to the cardinality of the set $\{(p, q) | r \in A(p, q)\}$ for all r. Let M_i be the marking such that $M_i(p) = \sum_q |A(p, q)|$. Note that as $|A(p, q)| \le |P|$ we have $M_i(p) \le f(v_p^i)$. Symmetrically, let M_f be the marking such that $M_f(q) = \sum_p |A(p, q)|$; we have $M_f(q) \le f(v_q^f, o)$. We are going to construct a history from M_i to M_m and from M_m to M_f.

Example 10. In our running example with $\mathcal{R}$, we obtain $M_i = \{PE, PE, R, R\}$, $M_f = \{E, E, P1, P1\}$, and $M_m = \{PE, E, R, P1\}$.

We build a history from M_i to M_m with trajectories labeled by allowed pairs (p, q) with many trajectories per pair. Each trajectory for pair (p, q) starts in place p. The stability condition guarantees that we can extend some trajectory to extend the set of places reached by trajectories labeled (p, q), until trajectories of every pair have reached all allowed intermediate places r such that (p, r, q) is allowed. For each reached place r some trajectory stays in r until the end of the history. The history from M_m to M_f is built in a similar way but using backward search from M_f. After combining the two histories into a history from M_i to M_f, we duplicate some trajectory for each pair of places until we have a history from M to M'. The construction consists of technical details and can be found in the full version.

Finally, we extract a firing sequence from the realizable history from M to M' by associating a transition and an iteration count to each step of the history. Each step with k trajectories going from p_s to p_d with $p_s \neq p_d$ is associated to a transition t iterated k times from p_s to p_d, where t realizes the step. This is possible by realizability of the history.

5.3   Correctness Given a Stable Restriction Set

We prove that given a stable set of correct restrictions, the algorithm always yields a correct answer in polynomial time. In case of a near-miss, both reporting the near-miss and correctly resolving reachability is considered a correct answer.

A near miss is reported in two cases of the solution flow construction, the second being more technical. We give a sketch of the proof, the technical details are provided in the full version.

Lemma 1. *The near-miss reports are correct.*

Proof (Sketch). We prove that the algorithm's reports of near-misses are correct for a net N, markings M, M' and a stable set of restrictions $\mathcal{R}$. A near miss is reported in two cases. In the first case we cannot decrease some edge capacity by $|P|$, after having attempted at most $|P| - 1$ decreases for this edge beforehand. This corresponds to a place of M or M' having more than 0 but less than $|P|^2$ tokens, which constitutes a near-miss.

In the second case after decreasing the capacity by $|P|$ along each of the b allowed pairs, there is some cut (V_I, V_O) with capacity less than $|M| - b|P|$. Each decrease operation decreases the capacity of each cut at most by $2|P|$, so the original capacity of the cut is less than $|M| + b|P|$. On the other hand, it is strictly more than $|M|$, as decreasing by $|P|$ along some pair reduced the capacity by more than $|P|$, which is impossible for any minimum cut by the second flow-based stability condition. The sets $X = V_I \cap \{v_p^i | p \in P\}$ and $Y = V_I \cap \{v_p^f | p \in P\}$ provide a near-miss.

If the algorithm does not report a near-miss, then it successfully constructs a solution flow and reports that M' is reachable from M. A realizable history can be constructed from the solution flow, proving that M can reach M'. Moreover the realizable history and then firing sequence from M to M' can be constructed in polynomial time and are correct by construction.

Lemma 2. *The algorithm runs in polynomial time given a stable set of restrictions.*

The runtime analysis is straightforward, and can be found in the full version.

5.4 Computing a Stable Restriction Set

We show that there is a polynomial algorithm that either computes a stable restriction set, or correctly reports unreachability. Starting with the empty set of restrictions, the algorithm repeatedly finds violations of the stability conditions and modifies the restriction set by adding some correct restrictions, or reports unreachability. Once no violations can be found, the algorithm terminates. As the total number of possible triples is $|P|^3$, only a polynomial number of iterations is needed. It remains to show that the violations as well as the corresponding additional correct restrictions can be found in polynomial time.

First Condition. A violation can be found by computing the maximum flow. Such a violation immediately implies unreachability, since a realizable history induces a maximum flow of value $|M|$.

Second Condition. A violation can be found by considering all the allowed pairs of places (p, q) and computing the maximum flow after decreasing the capacity by one along (p, q). If the decrease is successful and the maximum flow is $|M| - 2$, then (p, q) is a violating pair, as argued in the section with the flow-based stability conditions. We add new correct restrictions by forbidding it. If the decrease yields a maximum flow of $|M| - 1$ then this pair does not create a violation. If the decrease is not possible, then we add new correct restrictions by forbidding (p, q). Indeed if the decrease is not possible, then the capacity between i and v_p^i (resp. between v_q^f and o) is zero. The pair (p, q) must be forbidden as there is no realizable history in which a token goes from p to q. The pair provides a violation of the condition by the cut which puts only v_p^i and o into the outlet component V_O (resp., only v_q^f and i into the inlet component V_I) and which is minimal because it has capacity $|M|$.

Third and Fourth Condition. Checking for violations of reachability-based stability conditions shares part of the approach used to construct a history out of a solution flow. For the third condition, the algorithm enumerates upper bounds on an extended set of restrictions $\mathcal{R}'$ violating the condition. We start with $\mathcal{R}'$ equal to all the triples. We observe that $\mathcal{R}'$ cannot contain (p, p, q) for any pair (p, q) such that (p, p, q) is not in $\mathcal{R}$. We exclude such (p, p, q) from $\mathcal{R}'$. Then as long as there is a transition $s \xrightarrow{o} d$ and there are triples $(p, s, q), (p', o, q') \notin \mathcal{R}'$ and $(p, d, q) \in \mathcal{R}' \setminus \mathcal{R}$, we exclude (p, d, q) from $\mathcal{R}'$. If we end up proving that $\mathcal{R}' = \mathcal{R}$, there can be no violation.

Otherwise we prove that all the restrictions in $\mathcal{R}'$ are correct and thus that $\mathcal{R}$ is extendable. Indeed, by induction, any history satisfying the restrictions in $\mathcal{R}$ on all steps must also satisfy the restrictions in $\mathcal{R}'$.

The fourth condition is handled in a symmetric way.

Example 11. In our running example, starting from an empty restriction set, the second condition reports violations because decreasing is not possible. It forbids all the pairs but $(PE, E), (PE, P1), (R, E), (R, P1)$. Checking violations of the third condition forbids all triples except (PE, PE, E), (PE, E, E),

$(R, R, P1)$, $(R, P1, P1)$, $(R, P2, P1)$. Checking the fourth condition additionally forbids $(R, P2, P1)$ leaving only four allowed triples (PE, PE, E), (PE, E, E), $(R, R, P1)$, $(R, P1, P1)$. This set of restrictions is stable.

This procedure for constructing a stable set of restrictions, coupled with the previous algorithm in which the stable set was part of the input, completes the proof of Theorem 5.

6 Conclusion and Future Work

We have considered two restrictions of the IO net reachability problem with a promise for much simpler verification for some applications and established the reachability complexity in both these cases, which is NP-complete in one case and polynomial in the other.

We leave the question of complexity of set-set reachability under these restrictions for future research. Another related question is defining a notion of "approximate" reachability that would provide a reduction in complexity for IO nets, as merely bounding the maximum difference between token counts or the sum of differences preserves PSPACE-hardness of the reachability problem.

Acknowledgements. We wish to thank Javier Esparza for useful discussions. We are also grateful to the anonymous reviewers for their advice regarding the presentation.

References

1. Angeli, D., De Leenheer, P., Sontag, E.D.: A Petri net approach to the study of persistence in chemical reaction networks. Math. Biosci. **210**(2), 598–618 (2007)
2. Angluin, D., Aspnes, J., Eisenstat, D., Ruppert, E.: The computational power of population protocols. Distrib. Comput. **20**(4), 279–304 (2007)
3. Baldan, P., Cocco, N., Marin, A., Simeoni, M.: Petri nets for modelling metabolic pathways: a survey. Nat. Comput. **9**(4), 955–989 (2010)
4. Cardoza, E., Lipton, R.J., Meyer, A.R.: Exponential space complete problems for Petri nets and commutative semigroups: preliminary report. In: Chandra, A.K., Wotschke,D., Friedman, E.P., Harrison, M.A. (eds.) Proceedings of the 8th Annual ACM Symposium on Theory of Computing, Hershey, Pennsylvania, USA, 3–5 May 1976, pp. 50–54. ACM (1976)
5. Craciun, G., Tang, Y., Feinberg, M.: Understanding bistability in complex enzyme-driven reaction networks. In: Proceedings of the National Academy of Sciences of the United States of America (2006)
6. Czerwinski, W., Lasota, S., Lazic, R., Leroux, J., Mazowiecki, F.: The reachability problem for Petri nets is not elementary. In: Charikar, M., Cohen, E. (eds.) Proceedings of the 51st Annual ACM SIGACT Symposium on Theory of Computing, STOC 2019, Phoenix, AZ, USA, 23–26 June 2019, pp. 24–33. ACM (2019)
7. David, R., Alla, H.: Petri nets for modeling of dynamic systems: a survey. Automatica **30**(2), 175–202 (1994)
8. Dinits, E.A.: Algorithm for solution of a problem of maximum flow in a network with power estimation. Sov. Math. Dokl. **11**, 1277–1280 (1970)

9. Esparza, J., Raskin, M., Weil-Kennedy, C.: Parameterized analysis of immediate observation Petri nets. In: Donatelli, S., Haar, S. (eds.) PETRI NETS 2019. LNCS, vol. 11522, pp. 365–385. Springer, Cham (2019). https://doi.org/10.1007/978-3-030-21571-2_20

10. Ford, L.R., Fulkerson, D.R.: Maximal flow through a network. Can. J. Math. **8**, 399–404 (1956)

11. Lin, H.: Stratifying winning positions in parity games. In: van Hee, K.M., Valk, R. (eds.) PETRI NETS 2008. LNCS, vol. 5062, pp. 9–11. Springer, Heidelberg (2008). https://doi.org/10.1007/978-3-540-68746-7_4

12. Montanari, U., Rossi, F.: Contextual nets. Acta Informatica **32**(6), 545–596 (1995)

13. Raskin, M., Weil-Kennedy, C.: Efficient restrictions of immediate observation Petri nets. CoRR, abs/2007.09189 (2020)

14. Raskin, M., Weil-Kennedy, C., Esparza, J.: Flatness and complexity of immediate observation Petri nets. In: CONCUR 2020 (2020)

Binary Expression of Ancestors
in the Collatz Graph

Tristan Stérin[✉]

Hamilton Institute, Department of Computer Science, Maynooth University,
County Kildare, Ireland
tristan.sterin@mu.ie
https://dna.hamilton.ie/tsterin/

Abstract. The Collatz graph is a directed graph with natural number nodes and where there is an edge from node x to node $T(x) = T_0(x) = x/2$ if x is even, or to node $T(x) = T_1(x) = \frac{3x+1}{2}$ if x is odd. Studying the Collatz graph in binary reveals complex message passing behaviors based on carry propagation which seem to capture the essential dynamics and complexity of the Collatz process. We study the set $\mathcal{E}\mathrm{Pred}_k(x)$ that contains the binary expression of any ancestor y that reaches x with a limited budget of k applications of T_1. The set $\mathcal{E}\mathrm{Pred}_k(x)$ is known to be regular, Shallit and Wilson [EATCS 1992]. In this paper, we find that the structure of the Collatz graph naturally leads to the construction of a regular expression, $\mathbf{reg}_k(x)$, which defines $\mathcal{E}\mathrm{Pred}_k(x)$. Our construction, is exponential in k which improves upon the doubly exponentially construction of Shallit and Wilson. Furthermore, our result generalises Colussi's work on the $x = 1$ case [TCS 2011] to any natural number x, and gives mathematical and algorithmic (Code available here: https://github.com/tcosmo/coreli.) tools for further exploration of the Collatz graph in binary.

1 Introduction

Let $\mathbb{N} = \{0, 1, \dots\}$. The Collatz map, $T : \mathbb{N} \to \mathbb{N}$, is defined by $T(x) = T_0(x) = x/2$ if x is even or $T(x) = T_1(x) = (3x+1)/2$ if x is odd. The Collatz graph is the directed graph generated by T, nodes are all $x \in \mathbb{N}$ and arcs are $(x, T(x))$. This map, and its graph, have been widely studied (see surveys [16] and [17]) and research has been driven by a problem, open at least since the 60s: **the Collatz conjecture**. The conjecture states that, in the Collatz graph, any strictly positive natural number is a predecessor of 1. In other words, any $x > 0$ reaches 1 after a finite number of T-iterations. As of 2020, the Collatz conjecture has been tested for all natural numbers below 2^{68} without any counterexample found [1].

Research supported by European Research Council (ERC) under the European Union's Horizon 2020 research and innovation programme (grant agreement No 772766, Active-DNA project), and Science Foundation Ireland (SFI) under Grant number 18/ERCS/5746.

S. Schmitz and I. Potapov (Eds.): RP 2020, LNCS 12448, pp. 115–130, 2020.
https://doi.org/10.1007/978-3-030-61739-4_8

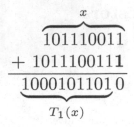

(a) The sum $3x + 1$ in binary. The number x gets added to $2x + 1$ which in binary is the left shift of x to which $\mathbf{1}$ is added.

$$\overline{0}0\overline{1}0\overline{1}\overline{1}10 0\overline{1}\overline{1}\overline{0}$$
$$\overline{10001011010}$$

(b) The sum $3x + 1$ interpreted as: "each bit of x sums with its right neighbour and the neighbour's potential carry". Carries are represented by dots. The $+1$ part of the operation is embedded in a carry on a fictional 0 to the right of the rightmost 1 bit. The first step, at the rightmost end, reads: $1 + \overline{0}$ which produces an ouput of 0 and transports the carry from to $\overline{1}$.

Fig. 1. Two ways to interpret the operation $3x+1$ in binary, illustrated on the number x with binary representation 101110011. The method shown in (b) highlights carry propagation.

There has been a fruitful trend of studying the Collatz process in binary [2–4,6,12,18,20]. That is because, the maps T_0 and T_1 have natural binary interpretations. The action of T_0 corresponds to shifting the input's binary representation to the right – deleting a trailing 0. While writing $T_1(x) = (3x + 1)/2 = (x + (2x + 1))/2$ reveals an interesting mechanism. In binary, the operation $x + (2x + 1)$ corresponds to adding x to its left-shifted version where the least significant bit has been set to 1, Fig. 1a. Equivalently $x + (2x + 1)$ corresponds to each bit of x being added to its right neighbour and the potential carry being placed on that neighbour. The $+1$ part of the operation can be represented as a carry appearing *ex nihilo* after the rightmost 1, Fig. 1b. The described mechanism results in the propagation of a carry within the binary representation of x: two consecutive 1s create a carry while two consecutive 0s absorb an incoming carry. Representing trajectories in the Collatz graph with the carry-annotated base 2 representation of Fig. 1b leads to complex, "Quasi Cellular Automaton" evolution diagrams which seem to encompass the overall complexity of the Collatz process [5]. These cary-annoted evolution diagrams are studied in depth in [21].

Here, we ask the following question: for a given bit string ω, what is the shape of the bit strings which "degrade" into ω under the action of the Collatz process? Said otherwise, for an arbitrary x, can we characterize the binary expansion of all y which reach x in the Collatz process? To answer that question, we find that it is natural to put a budget on the number of times the map T_1 is used (see Remark 26) and we study $\mathcal{E}\mathrm{Pred}_k(x)$ the set of binary expressions of all y which reach x by using the map T_1 exactly k times and the map T_0 an arbitrary number of times. There is a high-level argument which shows that for each x and k the set $\mathcal{E}\mathrm{Pred}_k(x)$ is regular [20]: the binary interpretation of the $3x + 1$ operation as shown in Fig. 1b can be performed by a 4-state, reversible, Finite

State Transducer (which states correspond to symbols $0, \bar{0}, 1, \bar{1}$). In [20], the authors make the point that having a budget of k on the map T_1 corresponds to iterating that transducer k times. Since finite iterations of FSTs lead to regular languages, $\mathcal{E}\mathrm{Pred}_k(x)$ is regular. However, while it gives regular structure to the set $\mathcal{E}\mathrm{Pred}_k(x)$, from the point of view of regular expressions, their argument does not lead to a tractable representation of $\mathcal{E}\mathrm{Pred}_k(x)$: they construct exponentially large FSTs in k, leading to doubly exponentially large regular expressions when using general purpose regular expression generation algorithm [11].

In this paper, we find that the knowledge about the binary structure of ancestors of x is embedded in the *geometry* of finite paths that reach x in the Collatz graph. By geometry of a path, we mean the *parity vector* [15,19,23,24] associated to that path, which corresponds to looking at the path's elements modulo 2 (Fig. 2). We find that there is a tight link between the shape of the parity vector and the binary expression of the first element on the path, which is an ancestor of x. Hence, we focus on characterizing the shapes of parity vectors of paths ending in x and then translate those shapes into binary expressions of ancestors. The budget of k applications of the map T_1 will translate to the constraint of having k 1s in the parity vectors we consider. Our main result exploits the mapping between constrained parity vectors and binary representation of ancestors at "T_1-distance" k, in order to construct a regular expression $\mathrm{reg}_k(x)$ which defines $\mathcal{E}\mathrm{Pred}_k(x)$. As the number of possible shapes of constrained parity vectors grows exponentially with k, these regular expressions are big[1] but only exponential in k (against doubly exponential in k in previous constructions [20]):

Theorem 1. *For all $x \in \mathbb{N}$, for all $k \in \mathbb{N}$ there exists a regular expression $\mathrm{reg}_k(x)$ that defines $\mathcal{E}\mathrm{Pred}_k(x)$. The regular expression $\mathrm{reg}_k(x)$ is structured as a tree with $2^k 3^{k(k-1)/2}$ branches, alphabetic width $O(2^k 3^{k(k+1)/2})$ and star height equal to 1.*

Our result generalises [6] which focused on the case $x = 1$. We claim that the framework we introduce is more general than [6,12] and that, in potential future work, it could easily be applied to generalisations of the Collatz map such as the T_q maps[2] [9]. Our result improves [20] by an exponential factor which makes our construction more fit for pratical use. We have implemented the construction of Theorem 1 (see Appendix C in the expanded version [22]) and claim that it gives a new exploratory tool for studying the Collatz process in binary. Indeed, for any x and k, we can sample $\mathrm{reg}_k(x)$ in order to analyse the different mechanisms by which the Collatz process transforms an input string into the binary representation of x in k odd steps. Also, from our result, one can also easily sample from $\mathrm{reg}_k(x)$ the *smallest* ancestor of x at T_1-distance k which suggests a new approach for future work in trying to understand how the Collatz process *optimally encodes* the structure of x in ancestors at T_1-distance k.

---

[1] Appendix D in [22] shows $\mathrm{reg}_4(1)$ which gives an idea of how large the regular expressions get.

[2] Defined, for q odd, by $T_q(x) = x/2$ if x is even or $T_q(x) = qx + 1$ if x is odd. These maps are as mysterious as the Collatz map.

In future work, we plan to use our algorithm as a tool to further understand the dynamics of the Collatz process in binary. In particular, we are very much concerned by the question: "Can the Collatz process compute?". Indeed, direct generalisations of the Collatz process are known to have full Turing power [7,13,14]. While the Collatz conjecture, by characterizing the long term behavior of any trajectory, seems to imply that there are some limitations on the computational power of the Collatz process, nothing is known. We believe that further studying carry propagation diagrams in the binary Collatz process can lead to answers on the computational power of the Collatz process and that, the tools built in this article can support that research.

2   Parity Vectors and Occurrences of Parity Vectors

Let $\mathbb{N} = \{0, 1, \dots\}$. We recall that the Collatz map $T : \mathbb{N} \to \mathbb{N}$, is defined by $T(x) = T_0(x) = x/2$ if x is even or $T(x) = T_1(x) = (3x + 1)/2$ if x is odd. The concept of parity vector was introduced in [23] (under the name *encoding vector*) and used, for instance, in [15,19,24]. While we work with the same concept, we introduce a slightly different representation[3] of parity vectors by using arrows $\downarrow$ and $\leftarrow$ instead of bits 0 and 1. In this Section, we introduce notation to manipulate *occurrences* of parity vectors in the Collatz graph and reformulate a crucial result of [24] in our framework (Theorem 8).

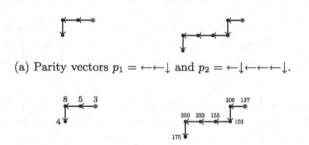

(a) Parity vectors $p_1 = \leftarrow\leftarrow\downarrow$ and $p_2 = \leftarrow\downarrow\leftarrow\leftarrow\leftarrow\downarrow$.

(b) An occurrence of $p_1 = \leftarrow\leftarrow\downarrow$ and an occurrence of $p_2 = \leftarrow\downarrow\leftarrow\leftarrow\leftarrow\downarrow$.

Fig. 2. Two parity vectors and one of their occurrences. In order to represent parity vectors, we use arrows $\downarrow$ and $\leftarrow$ instead of bits 0 and 1. When drawn in the plane, parity vectors read from **right** to **left**, start at the red dot. (Color figure online)

Definition 2 (Parity Vector). *A parity vector p is a word in $\{\downarrow, \leftarrow\}^*$, i.e. a finite word, possibly empty, over the alphabet $\{\downarrow, \leftarrow\}$. We call $\mathcal{P}$ the set of all parity vectors. The empty parity vector is ϵ. We define $\cdot$ to be the concatenation*

---

[3] This is done both because, in this format, parity vectors can be represented nicely in the plane (see Fig. 2), and because binary strings will be omnipresent in Sect. 3 and we don't want to confuse the reader with too many of them.

operation on parity vectors: $p = p_1 \cdot p_2$ *is the parity vector consisting of the arrows of p_1 followed by the arrows of p_2. We use exponentiation in its usual meaning:* $p^n = p \cdot p \ldots \cdot p$, n *times.*

Definition 3 (Norm and span). *As in [24], we define two useful metrics on parity vectors: (a) the norm of p, written $||p||$, is the total number of arrows in p and (b) the span[4] of p, written $l(p)$, is the number of arrows of type $\leftarrow$ in p.*

Definition 4 (Occurrence of a parity vector). *Let $p = a_0 \cdot \ldots \cdot a_{n-1} \in \mathcal{P}$ be a parity vector with $a_i \in \{\downarrow, \leftarrow\}$ and $n = ||p||$. An occurrence of p in the Collatz graph, or, for short, an occurrence of p, is a $(n+1)$-tuple, $(o_0, \ldots, o_{||p||}) \in \mathbb{N}^{||p||+1}$ such that, for $0 \leq i < ||p||$, $o_{i+1} = T_0(o_i)$ if $a_i = \downarrow$ or $o_{i+1} = T_1(o_i)$ if $a_i = \leftarrow$.*

Definition 5 (Set of occurrences of a parity vector: $\alpha(p)$). *Let $p \in \mathcal{P}$. We call $\alpha(p)$ the set of all the occurrences of the parity vector p. We order this set by the first number of each occurrence. Then, $\alpha_i(p) \in \mathbb{N}^{||p||+1}$ denotes the ith occurrence of p within that order and $\alpha_{i,j}(p)$, with $0 \leq j \leq ||p||$, denotes the jth term of the ith occurrence. In order to facilitate reading, we will write $\alpha_{i,-1}(p)$ instead of $\alpha_{i,||p||}(p)$ to refer to the last element of the occurrence $\alpha_i(p)$. If the context clearly states the parity vector p we will abuse notation and write $\alpha_{i,j}$ instead of $\alpha_{i,j}(p)$.*

Example 6. Figure 2a shows two parity vectors in $\mathcal{P}$: $p_1 = \leftarrow\leftarrow\downarrow$ and $p_2 = \leftarrow\downarrow\leftarrow\leftarrow\leftarrow\downarrow$. We have: $||p_1|| = 3$, $l(p_1) = 1$ and $||p_2|| = 6$, $l(p_2) = 4$. In Fig. 2b, it can be proved that we have $\alpha_0(p_1) = (3, 5, 8, 4)$ and $\alpha_2(p_2) = (137, 206, 103, 155, 233, 350, 175)$.

Definition 7 (Feasibility). *A parity vector $p \in \mathcal{P}$ is said to be feasible if it has at least one occurrence, i.e. if $\alpha_0(p)$ is defined.*

The question "Are all parity vectors feasible?" is answered positively in [24] (Lemma 3.1). This result is key to our work and we reformulate it in terms of occurrences of parity vectors:

Theorem 8 (All parity vectors are feasible). *Let $p \in \mathcal{P}$. Then:*

1. *p is feasible i.e. $\alpha_0 = (\alpha_{0,0}, \ldots, \alpha_{0,-1}) \in \mathbb{N}^{||p||+1}$ is defined*
2. *$\alpha_{0,0} < 2^{||p||}$ and $\alpha_{0,-1} < 3^{l(p)}$*
3. *Finally we can completely characterize $\alpha_{i,0}$ and $\alpha_{i,-1}$ with: $\alpha_{i,0} = 2^{||p||}i + \alpha_{0,0}$ and $\alpha_{i,-1} = 3^{l(p)}i + \alpha_{0,-1}$*

Proof. This Theorem is essentially a reformulation of Lemma 3.1 in [24]. We give the proof in [22], Appendix B. □

Example 9. Figure 3 in [22] illustrates the knowledge that Theorem 8 gives on the structure of $\alpha(p)$, the set of occurrences[5] of the parity vector p.

---

[4] Called *length* in [24]. We change terminology to avoid confusion with the notion of length of a word over an alphabet. However, we keep the same mathematical notation $l(p)$.

[5] The result of [19] implies that one can prove the Collatz conjecture by only proving it for $\alpha_{i,j}(p)$ for all $i \in \mathbb{N}$, for any $p \in \mathcal{P}$, for any $0 \leq j \leq ||p||$.

3   First Occurrence of Parity Vectors

In this Section, we show that there is a direct link between the $||p||$ arrows of a parity vector p and the $||p||$ bits of the binary representation – including potential leading 0s – of $\alpha_{0,0}(p)$ (Theorem 18). Then, we show that first occurrences of parity vectors can be arranged in a remarkably symmetric binary tree: the $(\alpha_{0,-1})$-tree (Theorem 22). As we work in binary, let's introduce some notation:

Definition 10 (The set $\mathcal{B}^*$). *Let $\mathcal{B}^*$ be the set of finite (possibly empty) words written on the alphabet $\mathcal{B} = \{0, 1\}$. The empty word, is denoted by η. We define $\bullet$, the concatenation operator on these words and we use exponentiation in its usual meaning. Finally, for $\omega \in \mathcal{B}^*$, $|\omega|$ refers to the length (number of symbols) in the binary word ω.*

Definition 11 (The interpretations[6] $\mathcal{I}$ and $\mathcal{I}_n^{-1}$). *Each word $\omega \in \mathcal{B}^*$ can, in a standard way, be interpreted as the binary representation of a number in $\mathbb{N}$. The function $\mathcal{I} : \mathcal{B}^* \to \mathbb{N}$ gives this interpretation. By convention, $\mathcal{I}(\eta) = 0$. Reciprocally, the partial function $\mathcal{I}_n^{-1} : \mathbb{N} \to \mathcal{B}_n^*$, where $\mathcal{B}_n^*$ is the set of $\omega \in \mathcal{B}^*$ with $|\omega| = n$, gives the binary representation of $x \in \mathbb{N}$ on n bits. The value of $\mathcal{I}_n^{-1}(x)$ is defined only when $n \geq \lfloor log_2(2x + 1) \rfloor$. We set $\mathcal{I}_0^{-1}(0) = \eta$. Finally, by $\mathcal{I}^{-1}(x)$ we refer to the binary representation of $x \in \mathbb{N}$ without any leading 0. Formally, $\mathcal{I}^{-1}(x) = \mathcal{I}_{\lfloor log_2(x) \rfloor +1}^{-1}(x)$ if $x \neq 0$ and $\mathcal{I}^{-1}(0) = \mathcal{I}_1^{-1}(0) = 0$.*

Example 12. $\mathcal{I}(11) = \mathcal{I}(0011) = 3$, $\mathcal{I}^{-1}(3) = \mathcal{I}_2^{-1}(3) = 11$ and $\mathcal{I}_7^{-1}(3) = 0000011$.

3.1   Constructing $\alpha_{0,0}$

Let's notice that we have the following bijection (similarly introduced in [23]):

Lemma 13. *Define $\mathcal{P}_n = \{p \in \mathcal{P}$ with $||p|| = n\}$. Then the function $f : \mathcal{P}_n \to \{0, \ldots, 2^n - 1\}$ defined by $f(p) = \alpha_{0,0}(p)$ is a bijection.*

Proof **(Sketch).** By cardinality, only injectivity is to prove which comes by determinism of the Collatz process. Full proof in the expanded version [22].   □

We can now define the *Collatz encoding* of a parity vector $p \in \mathcal{P}$:

Definition 14 (Collatz encoding of a parity vector p). *We define $\mathcal{E} : \mathcal{P} \to \mathcal{B}^*$ the Collatz encoding function of parity vectors to be: $\mathcal{E}(p) = \mathcal{I}_{||p||}^{-1}(\alpha_{0,0}(p))$. The function $\mathcal{E}$ is well defined since, by Theorem 8, $\alpha_{0,0}(p) < 2^{||p||}$. By Lemma 13, $\mathcal{E}$ is bijective hence $\mathcal{E}^{-1} : \mathcal{B}^* \to \mathcal{P}$ is naturally defined.*

Example 15. $\mathcal{E}(p)$ is the binary representation of $\alpha_{0,0}(p)$ on $||p||$ bits. We have: $\mathcal{E}(\downarrow\downarrow) = 00$ or $\mathcal{E}(\leftarrow\downarrow\downarrow) = 101$ (see Fig. 4 in [22]).

---

[6] We do not use the notation $\llbracket \cdot \rrbracket$ and its inverse $\llbracket \cdot \rrbracket^{-1}$ of [6,12] in order to avoid confusion. Indeed, in [6,12], the use of this notation is meant to preserve leading 0s while we crucially need to control them in order to define the encoding function $\mathcal{E}$.

Definition 16 (Admissibility of an arrow). *Let $a \in \{\downarrow, \leftarrow\}$. The arrow a is said to be admissible for the number x if and only if: ($a = \downarrow$ and x is even) or ($a = \leftarrow$ and x is odd).*

Lemma 17. *Let $p \in \mathcal{P}$ and $a \in \{\downarrow, \leftarrow\}$. Consider $\alpha_0(p \cdot a) = (\alpha_{0,0}(p \cdot a), \ldots, \alpha_{0,\|p \cdot a\|}(p \cdot a))$. Then two cases:*

- *If a is admissible for $\alpha_{0,-1}(p)$ then $(\alpha_{0,0}(p \cdot a), \ldots, \alpha_{0,\|p\|}(p \cdot a))$ is the first occurrence of p, i.e. we have: $\alpha_0(p) = (\alpha_{0,0}(p \cdot a), \ldots, \alpha_{0,\|p\|}(p \cdot a))$.*
- *If a is not admissible for $\alpha_{0,-1}(p)$ then $(\alpha_{0,0}(p \cdot a), \ldots, \alpha_{0,\|p\|}(p \cdot a))$ is the second occurrence of p, i.e. we have: $\alpha_1(p) = (\alpha_{0,0}(p \cdot a), \ldots, \alpha_{0,\|p\|}(p \cdot a))$.*

Proof. – If a is admissible for $\alpha_{0,-1}(p)$ then $p \cdot a$ is forward feasible for $\alpha_{0,0}(p)$ and $(\alpha_{0,0}(p), \ldots, \alpha_{0,-1}(p), T(\alpha_{0,-1}(p)))$ is an occurrence of $p \cdot a$. It has to be the first occurrence of $p \cdot a$ otherwise, the existence of a lower occurrence of $p \cdot a$ would contradict the fact that $\alpha_0(p) = (\alpha_{0,0}(p), \ldots, \alpha_{0,-1}(p))$ is the first occurrence of p.

- If a is not admissible for $\alpha_{0,-1}(p)$, consider $\alpha_0(p \cdot a) = (o_0, \ldots, o_{\|p\|+1})$ the first occurrence of $p \cdot a$. Then $(o_0, \ldots, o_{\|p\|})$ is an occurrence of p. It cannot be the first one since the first occurrence of p is followed by an arrow admissible for $\alpha_{0,-1}(p)$. However, by Theorem 8 we know that $o_0 = \alpha_{0,0}(p \cdot a) < 2^{\|p \cdot a\|} = 2^{\|p\|+1} = 2 * 2^{\|p\|}$. Thus we conclude that $(o_0, \ldots, o_{\|p\|})$ is the second occurrence of p, i.e. $o_0 = \alpha_{1,0}(p) = 2^{\|p\|} + \alpha_{0,0}(p)$ since for all $i \geq 2$, $\alpha_{i,0}(p) \geq 2 * 2^{\|p\|}$ by Theorem 8. □

Theorem 18 (Recursive structure of $\mathcal{E}$). *Let $n \in \mathbb{N}$. We have $\mathcal{E}(\epsilon) = \eta$. Then, for $p \in \mathcal{P}_n$ and $a \in \{\downarrow, \leftarrow\}$ we have $\mathcal{E}(p \cdot a) = 0 \bullet \mathcal{E}(p)$ if a is admissible for $\alpha_{0,-1}(p)$ and $\mathcal{E}(p \cdot a) = 1 \bullet \mathcal{E}(p)$ otherwise.*

Proof. By Definition 14, we have $\mathcal{E}(\epsilon) = \mathcal{I}_0^{-1}(\alpha_{0,0}(\epsilon)) = \mathcal{I}_0^{-1}(0)$ and $\mathcal{I}_0^{-1}(0) = \eta$ by Definition 11. Hence, $\mathcal{E}(\epsilon) = \eta$. Now, let $p \in \mathcal{P}_n$, $a \in \{\downarrow, \leftarrow\}$. Two cases:

- If a is admissible for $\alpha_{0,-1}(p)$, by Lemma 17 we have $\alpha_{0,0}(p \cdot a) = \alpha_{0,0}(p)$. Thus we get that $\mathcal{I}_{n+1}^{-1}(\alpha_{0,0}(p \cdot a)) = 0 \bullet \mathcal{I}_n^{-1}(\alpha_{0,0}(p))$ since prepending a 0 to a binary string doesn't change the number it represents. Hence, $\mathcal{E}(p \cdot a) = 0 \bullet \mathcal{E}(p)$.
- If a is not admissible for $\alpha_{0,-1}(p)$, by Lemma 17 and Theorem 8 we get $\alpha_{0,0}(p \cdot a) = \alpha_{1,0}(p) = 2^{\|p\|} + \alpha_{0,0}(p)$ which corresponds to prepending a bit 1 to the binary representation of $\alpha_{0,0}(p)$ on n bits. We conclude that $\mathcal{I}_{n+1}^{-1}(\alpha_{0,0}(p \cdot a)) = 1 \bullet \mathcal{I}_n^{-1}(\alpha_{0,0}(p))$. Hence, $\mathcal{E}(p \cdot a) = 1 \bullet \mathcal{E}(p)$.

□

Example 19. Figure 4 in [22] illustrates Theorem 18 on parity vectors of $\mathcal{P}_0, \mathcal{P}_1, \mathcal{P}_2, \mathcal{P}_3$.

3.2   Constructing $\alpha_{0,-1}$

Theorem 18 relies on knowing $\alpha_{0,-1}$ at each step in order to deduce the admissibility of the arrow which is being added. In this Section, we show that $\alpha_{0,-1}$ can also be recursively constructed. That construction will lead to a binary tree, the $(\alpha_{0,-1})$-tree in which each node corresponds to the first occurrence of a parity vector. The symmetries of this tree will be crucial to our main result, Theorem 1. The construction of $\alpha_{0,-1}$ relies on some elementary knowledge about groups of the form $\mathbb{Z}/3^k\mathbb{Z}$ and their multiplicative subgroup $(\mathbb{Z}/3^k\mathbb{Z})^*$. We recall the definition and main properties of these objects in Appendix A of [22]. In particular, we use the notation 2_k^{-1} to refer to the modular inverse of 2 in $\mathbb{Z}/3^k\mathbb{Z}$. Importantly, 2_k^{-1} is a primitive root of $(\mathbb{Z}/3^k\mathbb{Z})^*$. Those groups play an important role in our context because of the following result:

Lemma 20. *Let $p \in \mathcal{P}$. Then $l(p) \neq 0 \Leftrightarrow \alpha_{0,-1}(p) \in (\mathbb{Z}/3^{l(p)}\mathbb{Z})^*$. If $l(p) = 0$, $\alpha_{0,-1}(p) = 0$.*

Proof. We prove both directions: $\Rightarrow$: we suppose $l(p) \neq 0$. We know $\alpha_{0,-1}(p) < 3^{l(p)}$ (Theorem 8). We have to prove that $\alpha_{0,-1}(p)$ is not a multiple of three. The predecessor set of y, a multiple of 3, in the Collatz graph is reduced to $\{2^n y$ for $n \in \mathbb{N}\}$. Indeed, we know that all $2^n y$ are predecessors of y by the operator T_0. Furthermore, the operator $T_1^{-1}(y) = (2y - 1)/3$ never yields to an integer if inputed a multiple of three and all $2^n y$ are. Hence no parity vector p with $l(p) > 0$ can have an occurrence ending in a multiple of three and we have the result. $\Leftarrow$: if $l(p) = 0$ then $(\mathbb{Z}/3^{l(p)}\mathbb{Z})^* = \emptyset$ so we have the result. If $l(p) = 0$ then p has the form $p = (\downarrow)^n$. By Theorem 18, we deduce $\alpha_{0,0}(p) = 0$. Hence $\alpha_{0,-1}(p) = T^n(0) = 0$. $\square$

We can recursively construct $\alpha_{0,-1}$ with the analogous of T_0 and T_1 in $\mathbb{Z}/3^k\mathbb{Z}$:

Definition 21 ($T_{0,k}$ and $T_{1,k}$). *The functions $T_{0,k} : \mathbb{Z}/3^k\mathbb{Z} \to \mathbb{Z}/3^k\mathbb{Z}$ and $T_{1,k} : \mathbb{Z}/3^k\mathbb{Z} \to \mathbb{Z}/3^k\mathbb{Z}$ are defined by: $T_{0,k}(x) = 2_k^{-1}x$ and $T_{1,k}(x) = 2_k^{-1}(3x + 1)$.*

Theorem 22 (Recursive structure of $\alpha_{0,-1}$).   *Let $n \in \mathbb{N}$. We have $\alpha_{0,-1}(\epsilon) = 0$. Then, for some $p \in \mathcal{P}_n$ and $k = l(p)$ we have $\alpha_{0,-1}(p \cdot \downarrow) = T_{0,k}(\alpha_{0,-1}(p))$ and $\alpha_{0,-1}(p \cdot \leftarrow) = T_{1,k+1}(\alpha_{0,-1}(p))$.*

Proof **(Sketch).** By induction, using that $T_{0,k}(x) = (3^k + x)/2$ when x is odd and $T_{1,k+1}(x) = (3^k + 3x + 1)/2$ when x is even. Full proof in [22]. $\square$

Example 23. On Fig. 4 in [22], we are reading $\alpha_{0,-1}(\downarrow \leftarrow \leftarrow) = 8$. On the other hand, Theorem 22 claims that $\alpha_{0,-1}(\downarrow \leftarrow \leftarrow) = T_{1,2}(\alpha_{0,-1}(\downarrow \leftarrow)) = T_{1,2}(2)$. Let's verify that: $T_{1,2}(2) = 2_2^{-1}(3*2+1) = 3+2_2^{-1} = 3+\frac{3^2+1}{2} = 3+5 = 8$ as expected.

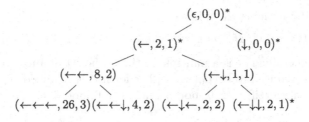

Fig. 3. First 4 levels of the $(\alpha_{0,-1})$-tree. Two symmetries are highlighted by $*$ and $\star$.

3.3   The $(\alpha_{0,-1})$-tree

Theorem 22 implies that the operators $T_{0,k}$ and $T_{1,k}$ naturally give birth to a binary tree ruling the construction of $\alpha_{0,-1}$. We call this tree the $(\alpha_{0,-1})$-tree:

Definition 24 (The $(\alpha_{0,-1})$-tree). *We call the $(\alpha_{0,-1})$-tree the binary tree with nodes in $\mathcal{N} \subset (\mathcal{P} \times \mathbb{N} \times \mathbb{N})$ constructed as follow, starting from node $x = (\epsilon, 0, 0)$:*

1. *The right child of (p, x, k) is $((p \cdot \downarrow), T_{0,k}(x), k)$*
2. *The left child of (p, x, k) is given by $((p \cdot \leftarrow), T_{1,k+1}(x), k + 1)$*

Lemma 25. *Nodes of the $(\alpha_{0,-1})$-tree are: $\mathcal{N} = \{(p, \alpha_{0,-1}(p), l(p)) \text{ for } p \in \mathcal{P}\}$.*

Proof. Each node of $(\alpha_{0,-1})$-tree corresponds to a first occurrence, immediate from Definition 24 and Theorem 22.   □

Symmetries of the $(\alpha_{0,-1})$-tree. Figure 3 illustrates the first four levels of the $(\alpha_{0,-1})$-tree. By construction of the $(\alpha_{0,-1})$-tree, if two nodes (p, x, k) and (p', x, k) share the same x and k they will be the root of very similar subtrees. This phenomenon is highlighted with the nodes $(\epsilon, 0, 0)$ and $(\downarrow, 0, 0)$, Fig. 3 doesn't show the sub-tree under $(\downarrow, 0, 0)$ as it can be entirely deduced from the sub-tree under $(\epsilon, 0, 0)$. The same would apply for the sub-trees under $(\leftarrow, 2, 1)$ and $(\leftarrow\downarrow\downarrow, 2, 1)$. These symmetries are closely related to the fact that keeping adding $\downarrow$ to a parity vector of span k will periodically enumerate $(\mathbb{Z}/3^k\mathbb{Z})^*$ (Lemma 36).

4   Regular Expressions Defining Ancestors Sets

Pursuing our primary goal, we wish to characterize the binary expression of ancestors of an arbitrary x in the Collatz graph. We decompose the set of all ancestors of x as the union on k of sets $\text{Pred}_k(x)$. The set $\text{Pred}_k(x)$ contains all the ancestors of x which use the map T_1 exactly k times in order to reach x – the map T_0 can be used an arbitrary number of times.

Remark 26. The set $\text{Pred}_k(x)$ appears naturally in a fast-forwarded version of the Collatz process where even steps are ignored and only odd steps are considered (see [6]). In the graph of that process, $\text{Pred}_k(x)$ corresponds to the set of ancestors of x at distance k.

Let's start by noticing the following:

Lemma 27. *Let $x \in \mathbb{N}$. If x is a multiple of 3 then: $\forall k > 0$, $Pred_k(x) = \emptyset$.*

Proof. More generally, if x is a multiple of three, the set of ancestors of x in the Collatz graph is reduced to $Pred_0(x) = \{2^n x \text{ for } n \in \mathbb{N}\}$. Indeed, $T_1^{-1}(x) = \frac{2x-1}{3}$ cannot be an integer if $x \equiv 0 \bmod 3$ and $x \equiv 0 \bmod 3 \Rightarrow \forall n \in \mathbb{N}$, $2^n x \equiv 0 \bmod 3$. $\square$

Remark 28. In fact, sets $Pred_k(x)$ are infinite for all k as soon as x is not a multiple of 3.

Thanks to Sect. 3, we know that we can describe the binary expression of elements of $Pred_k(x)$ by focusing on parity vectors: the function $\mathcal{E}$ will translate parity vectors to the binary expressions of ancestors. Let's make that link formal:

Definition 29 ($\mathcal{E}Pred_k(x)$). *Let $x \in \mathbb{N}$ and $k \in \mathbb{N}$. We define the set $\mathcal{E}Pred_k(x) \subset \mathcal{B}^*$ to be: $\mathcal{E}Pred_k(x) = \{\omega \bullet \mathcal{E}(p) \mid p \in \mathcal{P} \text{ such that } \alpha_{0,-1}(p) = x \bmod 3^k \text{ and } l(p) = k\}$. With $\omega = \eta$ if $x < 3^k$ or $\omega = \mathcal{I}^{-1}(i)$ otherwise, and $i = \lfloor \frac{x}{3^k} \rfloor$. By $x \bmod 3^k$, we mean "the rest in the Euclidean division of x by 3^k".*

The set $\mathcal{E}Pred_k(x)$ constains binary representations of elements of $Pred_k(x)$ – with potential leading 0s – in a one-to-one correspondence:

Lemma 30. *Let $x > 0$ and $k \in \mathbb{N}$. The sets $\mathcal{E}Pred_k(x)$ and $Pred_k(x)$ are in bijection by the function $g : \mathcal{E}Pred_k(x) \to Pred_k(x)$ defined by $g(\omega) = \mathcal{I}(\omega)$.*

Proof. Straightforward, proof in the expanded version [22]. $\square$

Hence, in order to describe $\mathcal{E}Pred_k(x)$ we are concerned by characterizing parity vectors p such that $\alpha_{0,-1}(p) = x \bmod 3^k$ and $l(p) = k$. Such p correspond to the symmetries that we highlighted in the $(\alpha_{0,-1})$-tree, they form an equivalence class of "k-span equivalence":

Definition 31 (k-span equivalence). *Two parity vectors $p_1, p_2 \in \mathcal{P}$ are said to be k-span equivalent if $l(p_1) = l(p_2) = k$ and $\alpha_{0,-1}(p_1) = \alpha_{0,-1}(p_2)$. We write $p_1 \simeq_k p_2$. Note that $\simeq_k$ is an equivalence relation.*

The following set of binary strings will play a central role in how we can describe k-span equivalence classes:

Definition 32 (Parity sequence of $(\mathbb{Z}/3^k\mathbb{Z})^*$). *For $k > 0$, we define $\Pi_k \in \mathcal{B}^*$, the parity sequence of $(\mathbb{Z}/3^k\mathbb{Z})^*$ as follows: $\Pi_k = b_0 \dots b_{\pi_k-1}$ with $|\Pi_k| = \pi_k = |(\mathbb{Z}/3^k\mathbb{Z})^*| = 2*3^{k-1}$ and, $b_{\pi_k-1-i} = 0$ if 2_k^{-i} is even and $b_{\pi_k-1-i} = 1$ if 2_k^{-i} is odd. By convention, we fix $\pi_0 = 1$.*

Example 33. For $k = 3$, we have $2_3^{-1} = 14$. The sequence of powers of 2_3^{-1} in $(\mathbb{Z}/3^k\mathbb{Z})^*$ is: $[1, 14, 7, 17, 22, 11, 19, 23, 25, 26, 13, 20, 10, 5, 16, 8, 4, 2]$. The associated parity sequence (0 when even and 1 when odd) is: 101101111010010000. Finally, Π_3 is the mirror image of this: $\Pi_3 = 000010010111101101$. We have: $\Pi_1 = 01$, $\Pi_2 = 000111$, $\Pi_3 = 000010010111101101$ and $\Pi_4 = 000000110010100100010110000111111001101011011101001111$.

Remark 34. The strings Π_k, or "seeds" in [6], have been studied in great depth in [18]. The author find that their structure is extremely complex, that they have numerous properties and that they can be defined in a lot of different ways. For instance, [12] uses the fact that strings Π_k correspond to the repetend of $1/3^k$ in binary.

Definition 35 (Rotation operator $\mathcal{R}_i(\cdot)$). *Let $\omega \in \mathcal{B}^*$ with $|\omega| = n$. Then, for $0 \le i < n$, $\mathcal{R}_i(\omega)$ denotes the ith rotation (or circular shift) to the right of ω. For instance, we have $\mathcal{R}_2(000111) = 110001$.*

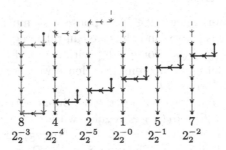

$$
\begin{array}{cccccc}
8 & 4 & 2 & 1 & 5 & 7 \\
2_2^{-3} & 2_2^{-4} & 2_2^{-5} & 2_2^{-0} & 2_2^{-1} & 2_2^{-2}
\end{array}
$$

Fig. 4. Illustration of Lemma 36. How the parity vector $p = \downarrow \leftarrow \leftarrow$ (in blue), with $l(p) = 2$, distributes on the elements of $(\mathbb{Z}/3^2\mathbb{Z})^*$. The first of occurrence of p is such that $\alpha_{0,-1} = 8 = 2_2^{-i_0} = 2_2^{-3}$. The parity vector p is k-span equivalent to the parity vector $p' = \downarrow \leftarrow \leftarrow (\downarrow)^6$ (in brown). (Color figure online)

From any parity vector p, we can create an infinite family of distinct parity vectors which are k-span equivalent to p:

Lemma 36. *Let $p \in \mathcal{P}$ and $k = l(p) > 0$. Define $p_n = p \cdot (\downarrow)^{n\pi_k}$, i.e. the parity vector p followed by $n\pi_k$ arrows of type $\downarrow$, where $\pi_k = |\Pi_k|$. Then, for all $n \in \mathbb{N}$ we have $p \simeq_k p_n$. Furthermore we can characterize $\alpha_{0,0}(p_n)$ through $\mathcal{E}(p_n)$ with:*

$$\mathcal{E}(p_{n+1}) = \mathcal{R}_{i_0}(\Pi_k) \bullet \mathcal{E}(p_n) \Leftrightarrow \mathcal{E}(p_n) = (\mathcal{R}_{i_0}(\Pi_k))^n \bullet \mathcal{E}(p)$$

With $0 \le i_0 < \pi_k$ such that $\alpha_{0,-1}(p) = 2_k^{-i_0}$ in $(\mathbb{Z}/3^k\mathbb{Z})^*$.

Proof **(Sketch).** Direct consequence of Theorem 18 and 22. Full proof in the expanded version [22]. □

Remark 37. The result of Lemma 36 is illustrated in Fig. 4. The parity vector p (in blue) distributes in a "spiral" around the elements of $(\mathbb{Z}/3^k\mathbb{Z})^*$. When π_k arrows of type $\downarrow$ have been added to p, a full "turn" has been done and we get a path k-span equivalent to p. As a consequence, following only right children in the $(\alpha_{0,-1})$-tree exhibits periods of length π_k which enumerate elements of $(\mathbb{Z}/3^k\mathbb{Z})^*$.

We now have all the element in order to characterize $\mathcal{E}\mathrm{Pred}_k(x)$ using regular expressions:

Theorem 1. *For all $x \in \mathbb{N}$, for all $k \in \mathbb{N}$ there exists a regular expression $\mathbf{reg}_k(x)$ that defines $\mathcal{E}\mathrm{Pred}_k(x)$. The regular expression $\mathbf{reg}_k(x)$ is structured as a tree with $2^k 3^{k(k-1)/2}$ branches, alphabetic width $O(2^k 3^{k(k+1)/2})$ and star height equal to 1.*

Proof. We are going to explicitly construct $\mathbf{reg}_k(x)$, a regular expression[7] which defines $\mathcal{E}\mathrm{Pred}_k(x)$. With the following preliminary argument we show that it is enough to construct $\mathbf{reg}_k(x)$ when x is not a multiple of 3 and $x < 3^k$. In other words, when $x \in (\mathbb{Z}/3^k\mathbb{Z})^*$.

Preliminary Argument. Let $x, k \in \mathbb{N}$. Suppose x is a multiple of 3. If $k > 0$, by Lemma 27, $\mathcal{E}\mathrm{Pred}_k(x)$ is empty and thus we can take $\mathbf{reg}_k(x) = \emptyset$ in that case. If $k = 0$, $\mathcal{E}\mathrm{Pred}_0(x) = \{\omega \bullet (0)^n \text{ for } n \in \mathbb{N}\}$ with $\omega = \eta$ if $x = 0$ or $\omega = \mathcal{I}^{-1}(x)$ otherwise (Definition 29 and Theorem 18). Hence we take $\mathbf{reg}_0(0) = (0)^*$ and $\mathbf{reg}_0(x) = (\mathcal{I}^{-1}(x))(0)^*$ for $x > 0$.

Suppose x is not a multiple of 3 and $x \geq 3^k$. Suppose that $\mathbf{reg}_k((x \bmod 3^k))$ exists, i.e. that the set $\mathcal{E}\mathrm{Pred}_k(x')$ is regular with $x' = (x \bmod 3^k)$. Then by Definition 29 we can take $\mathbf{reg}_k(x) = (\mathcal{I}^{-1}(\lfloor \frac{x}{3^k} \rfloor))(\mathbf{reg}_k((x \bmod 3^k)))$ in order to define $\mathcal{E}\mathrm{Pred}_k(x)$. Indeed, by Theorem 8, $\{p \in \mathcal{P} \mid \alpha_{i,-1}(p) = x\} = \{p \in \mathcal{P} \mid \alpha_{0,-1} = (x \bmod 3^k)\}$ with $i = \lfloor \frac{x}{3^k} \rfloor$.

Hence we just have to prove that $\mathbf{reg}_k(x)$ exists for all x, non multiple of three such that $x < 3^k$, i.e. $x \in (\mathbb{Z}/3^k\mathbb{Z})^*$. We prove by induction on k the following result:

$$H(k) = \text{``} \forall x \in (\mathbb{Z}/3^k\mathbb{Z})^* \text{ there exists } \mathbf{reg}_k(x) \text{ which defines } \mathcal{E}\mathrm{Pred}_k(x)\text{''}$$

Induction

Base step $k = 0$. Trivially true because $(\mathbb{Z}/3^k\mathbb{Z})^* = \emptyset$. Note that the following induction step will rely on knowing $\mathbf{reg}_0(0)$. We have shown above that $\mathbf{reg}_0(0) = (0)^*$.

Inductive Step. Let $k \in \mathbb{N}$ such that $H(k)$ holds. We show that $H(k + 1)$ holds. Let $x \in (\mathbb{Z}/3^{k+1}\mathbb{Z})^*$ and $0 \leq i_0 < \pi_{k+1}$ such that $x = 2_{k+1}^{-i_0}$. By Definition 29, in this case, we have $\mathcal{E}\mathrm{Pred}_{k+1}(x) = \{\mathcal{E}(p) \mid p \in \mathcal{P} \text{ such that } \alpha_{0,-1}(p) = x \text{ and } l(p) = k + 1\}$. Hence, characterizing $\mathcal{E}\mathrm{Pred}_{k+1}(x)$ boils down to characterizing the $(k + 1)$-span equivalence class: $\{p \mid p \in \mathcal{P} \text{ such that } \alpha_{0,-1}(p) = x \text{ and } l(p) = k + 1\} = \{p \mid (p, x, k + 1) \in \mathcal{N}\}$ (Lemma 25).

Hence, we take p such that $(p, x, k + 1)$ is in the $(\alpha_{0,-1})$-tree and we analyse its structure. To do so, we consider the surrounding of p in the $(\alpha_{0,-1})$-tree. This

---

[7] The regular expressions we work with are defined by the following BNF:

$$\text{reg} := \emptyset \mid (\omega \in \mathcal{B}^*) \mid (\text{reg}_1 | \text{reg}_2) \mid (\text{reg})^* \mid (\text{reg}_1)(\text{reg}_2)$$

For instance, the expression $(01)^*((00)|(11))$ matches any word of the form $(01)^n 00$ or $(01)^n 11$. We might omit some parenthesis when they are redundant.

will lead us to Eq. (1) which relates $\mathcal{E}(p)$ to the induction hypothesis. We are going to deploy Points 1, 2, 3, in order to show that the node $(p, x, k+1)$ can always be expressed in the context of Fig. 5:

In order to prove the generality of this situation, three points:

1. Since $l(p) = k + 1 \geq 1$ we can decompose $p = p_2 \cdot \leftarrow \cdot (\downarrow)^m$ with $m \in \mathbb{N}$ and $p_2 \in \mathcal{P}$ such that $l(p_2) = k$. We can write $m = n\pi_{k+1} + r$ with $r < \pi_{k+1}$ and $p = p_2 \cdot \leftarrow \cdot (\downarrow)^r \cdot (\downarrow)^{n\pi_{k+1}}$. We call the number n the *repeating value*. By Lemma 36, we know that $p \simeq_{k+1} p_2 \cdot \leftarrow \cdot (\downarrow)^r$. Hence, with $p' = p_2 \cdot \leftarrow \cdot (\downarrow)^r$, we have $\alpha_{0,-1}(p') = \alpha_{0,-1}(p) = x$ and $\mathcal{E}(p) = (\mathcal{R}_{i_0}(\Pi_{k+1}))^n \bullet \mathcal{E}(p')$. It remains to characterize $\mathcal{E}(p') = \mathcal{E}(p_2 \cdot \leftarrow \cdot (\downarrow)^r) = \mathcal{E}(p_1 \cdot (\downarrow)^r)$ with $p_1 = p_2 \cdot \leftarrow$.

2. Let's consider now $x_1 = \alpha_{0,-1}(p_2 \cdot \leftarrow)$. By Theorem 22, we have $x_1 = T_{1,k+1}(x_2) = T_{1,k+1}(2_k^{-i_2})$. Furthermore, by the same Theorem 22, we have $x = \alpha_{0,-1}(p') = \alpha_{0,-1}(p_1 \cdot \downarrow^r) = T_{0,k+1}^r(\alpha_{0,-1}(p_2 \cdot \leftarrow)) = T_{0,k+1}^r(x_1)$. Hence $x = 2_{k+1}^{-r} x_1$ and thus $x_1 = 2_{k+1}^{-i_1}$ with $0 \leq i_1 = -i_0 + r < \pi_{k+1}$. By Theorem 18, we deduce that: $\mathcal{E}(p' = p_1 \cdot (\downarrow)^r) = \omega \bullet \mathcal{E}(p_1)$. With $\omega = j_0 \ldots j_{r-1} \in \mathcal{B}^*$,

$$|\omega| = r, \text{ and } j_{r-i} = \begin{cases} 0 \text{ if } 2_{k+1}^{-i_1-i} \text{ is even} \\ 1 \text{ if } 2_{k+1}^{-i_1-i} \text{ is odd} \end{cases} \quad \text{with } 0 \leq i < r.$$

We refer to such ω by $\mathtt{join}_{i_2}$ because[8] it is uniquely determined by i_2 such that $x_1 = T_{1,k+1}(x_2) = T_{1,k+1}(2_k^{-i_2})$. Indeed, Lemma 44 of [22] shows that $T_{1,k+1}$ is injective on $\mathbb{Z}/3^k\mathbb{Z}$ hence $i_2 \neq i_2' \Rightarrow T_{1,k+1}(2_k^{-i_2}) \neq T_{1,k+1}(2_k^{-i_2'})$. Different values of i_2 will yield to different x_1 and thus different i_1, r and $\mathtt{join}_{i_2}$.

3. Let's consider $x_2 = \alpha_{0,-1}(p_2)$. If $k \neq 0$, by Lemma 20, we know that $x_2 \in (\mathbb{Z}/3^k\mathbb{Z})^*$ and we can write $x_2 = 2_k^{-i_2}$ with $0 \leq i_2 < \pi_k$. Note that if $k = 0$, by the same Lemma, we have $x_2 = 0 = 2_0^0$ by convention. Thus in all case we can write $x_2 = 2_k^{-i_2}$ with $0 \leq i_2 < \pi_k$. By Theorem 18, we deduce that $\mathcal{E}(p_1) = b_{i_2} \bullet \mathcal{E}(p_2)$ where $b_{i_2} = 0$ if $\leftarrow$ is admissble for x_2 and $b_{i_2} = 1$ otherwise. In other words: $b_{i_2} = \begin{cases} 0 \text{ if } 2_k^{-i_2} \text{ is odd} \\ 1 \text{ if } 2_k^{-i_2} \text{ is even} \end{cases}$.

Over all, from Points 1, 2, 3, we deduce that:

$$\mathcal{E}(p) = (\mathcal{R}_{i_0}(\Pi_{k+1}))^n \bullet (\mathtt{join}_{i_2}) \bullet (b_{i_2}) \bullet \mathcal{E}(p_2) \tag{1}$$

We have $l(p_2) = k$. If $k \neq 0$, we will be able to reduce to the induction hypothesis since $x_2 \in (\mathbb{Z}/3^k\mathbb{Z})^*$. If $k = 0$ we have $x_2 = 0$ and we use $\mathtt{reg}_0(0)$ previously constructed.

As a synthesis, notice that any value of $0 \leq i_2 < \pi_k$, any node $(p_2, x_2 = 2_k^{i_2}, k)$ and any repeating value $n \in \mathbb{N}$ will lead to the construction of a $(p, x, k+1)$ with a different p for each choice of i_2, p_2 and n. Hence, we have completely

---

[8] The name $\mathtt{join}$ refers to the fact that this parity sequence ω arises from "joining", in the $(\alpha_{0,-1})$-tree, $x_1 = 2_{k+1}^{i_1}$ to $x = 2_{k+1}^{i_0}$ with $r = i_0 - i_1 \geq 0$ arrows of type $\downarrow$. Notice that we can have $\mathtt{join}_{i_2} = \eta$ in the case where $r = 0$.

characterized the structure of nodes of the form $(p, x, k + 1)$. From the above analysis, we can deduce the recursive expression of $\mathbf{reg}_{k+1}(x)$, we have:

$$
\begin{aligned}
\mathbf{reg}_{k+1}(x) = (\mathcal{R}_{i_0}(\varPi_{k+1}))^* (\quad &(\mathtt{join}_0)\,(b_0)\,(\mathbf{reg}_k(2_k^{-0})) \quad | \\
&(\mathtt{join}_1)\,(b_1)\,(\mathbf{reg}_k(2_k^{-1})) \quad | \\
&(\mathtt{join}_2)\,(b_2)\,(\mathbf{reg}_k(2_k^{-2})) \quad | \\
&\qquad\qquad\vdots \\
&(\mathtt{join}_{\pi_k-1})\,(b_{\pi_k-1})\,(\mathbf{reg}_k(2_k^{-(\pi_k-1)})) \quad)
\end{aligned}
\tag{2}
$$

Note the amusing fact that for $k > 0$ the word $b_{\pi_k-1}b_{\pi_k-2}\ldots b_0$ is the binary complement of $\varPi_k$. The fact that $\mathbf{reg}_k(x)$ is structured as a tree is made obvious by Eq. (2), at each level $l \le k$ the branching factor is π_l. The number of branches is given by $\prod_{l=0}^{k} \pi_k = 2^k 3^{\frac{k(k-1)}{2}}$. The number of $\{0,1\}$ symbols on each branch is bounded by $2\sum_{l=0}^{k} \pi_k = O(3^k)$, hence, the alphabetic width of $\mathbf{reg}_k(x)$ which is the total number of $\{0,1\}$ symbols is $O(2^k 3^{\frac{k(k+1)}{2}})$. Finally, the star height[9][11] is 1 as directly derived from (2). $\square$

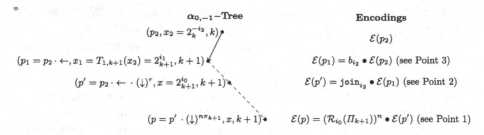

Fig. 5. Situation of the node $(p, x, k + 1)$ is the $(\alpha_{0,-1})$-tree. Each black dot is a node in the $(\alpha_{0,-1})$-tree. The solid edge reaches one left child while dashed edges represent variable numbers of right children (see Definition 24). This Figure is to be read bottom to top together with Points 1,2,3.

Remark 38. We remark that the number of branches of $\mathbf{reg}_k(x)$ corresponds to the size of "Level $k + 1$" that was computed in [10]. The author of [10] also remarks that this number corresponds to the number of different antisymmetric binary relations on a set of $k + 1$ labeled points [8].

Remark 39. The Collatz conjecture is equivalent to: for all x, there is k such that $x \in \mathrm{Pred}_k(1)$. Which means: for all x, there is k and n such that $\mathbf{reg}_k(1)$ matches

---

[9] The star height metric `height` is defined by $\mathtt{height}(s) = 0$ for $s \in \{\emptyset, \epsilon, 0, 1\}$, $\mathtt{height}(s_1|s_2) = \mathtt{height}(s_1 s_2) = \max(\mathtt{height}(s_1), \mathtt{height}(s_2))$ and $\mathtt{height}(s^*) = \mathtt{height}(s) + 1$.

$0^n \mathcal{I}^{-1}(x)$. Because the number of leading 0s in Π_k is equal to $\lfloor k \cdot \ln(3)/\ln(2) \rfloor$ (see [18]), we can bound the number of leading 0s that is accepted by $\text{reg}_k(1)$ and so we can bound n which is the number 0s to prepend to $\mathcal{I}^{-1}(x)$. For instance $n < (k+1)^2$ is a bound that works.[10]

Acknowledgement. Many thanks to Jose Capco, Damien Woods, Pierre-Étienne Meunier and Turlough Neary for their kind help, interest and feedback on this project. We also thank Jeffrey C. Lagarias for his surveys on the Collatz problem ([16] and [17]). We thank the OEIS, always of great help. Finally, sincere thanks to the anonymous reviewers. Their comments were very helpful, for instance making us realise the exponential gain of our construction compared to previous literature.

References

1. Barina, D.: Convergence verification of the Collatz problem. J. Supercomputing (2020). https://doi.org/10.1007/s11227-020-03368-x
2. Blazewicz, J., Pettorossi, A.: Some properties of binary sequences useful for proving Collatz's conjecture. Found. Control Eng. **8**, 53–63 (1983)
3. Bruschi, M.: Two cellular automata for the 3x+1 map (2005)
4. Capco, J.: Odd Collatz sequence and binary representations, preprint, March 2019. https://hal.archives-ouvertes.fr/hal-02062503
5. Cloney, T., Goles, E., Vichniac, G.Y.: The 3x+1 problem: a quasi cellular automaton. Complex Syst. **1**(2), 349–360 (1987)
6. Colussi, L.: The convergence classes of Collatz function. Theoret. Comput. Sci. (2011). https://doi.org/10.1016/j.tcs.2011.05.056
7. Conway, J.: Unpredictable iterations. In: Number Theory Conference (1972)
8. Foundation, O.: The Online Encyclopedia of Integer Sequences. A083667 (2020)
9. Franco, Z., Pomerance, C.: On a conjecture of Crandall concerning the $qx + 1$ problem. Math. Comput. **64**(211), 1333–1336 (1995). http://www.jstor.org/stable/2153499
10. Goodwin, J.R.: The 3x+1 problem and integer representations (2015)
11. Gruber, H., Holzer, M.: From finite automata to regular expressions and back-a summary on descriptional complexity. Electron. Proc. Theoret. Comput. Sci. 151 (2014). https://doi.org/10.4204/EPTCS.151.2
12. Hew, P.C.: Working in binary protects the repetends of $1/3^h$: comment on Colussi's 'The convergence classes of Collatz function'. Theoret. Comput. Sci. **618**, 135–141 (2016). https://doi.org/10.1016/j.tcs.2015.12.033
13. Koiran, P., Moore, C.: Closed-form analytic maps in one and two dimensions can simulate universal Turing machines. Theoret. Comput. Sci. **210**(1), 217–223 (1999). https://doi.org/10.1016/s0304-3975(98)00117-0
14. Kurtz, S.A., Simon, J.: The undecidability of the generalized Collatz problem. TAMC **2007**, 542–553 (2007). https://doi.org/10.1007/978-3-540-72504-6_49
15. Lagarias, J.C.: The 3x + 1 problem and its generalizations. Am. Math. Monthly **92**(1), 3–23 (1985). http://www.jstor.org/stable/2322189
16. Lagarias, J.C.: The 3x+1 problem: an annotated bibliography (1963–1999) (sorted by author) (2003)

---

[10] http://oeis.org/.

17. Lagarias, J.C.: The 3x+1 problem: an annotated bibliography, ii (2000–2009) (2006)
18. López, J., Stoll, P.: The 2-adic, binary and decimal periods of 1/3k approach full complexity for increasing k. Integers [electronic only] 5 (2012). https://doi.org/10.1515/integers-2012-0013
19. Monks, K.: The sufficiency of arithmetic progressions for the 3x + 1 conjecture. Proc. Am. Math. Soc. **134** (2006). https://doi.org/10.2307/4098142
20. Shallit, J., Wilson, D.A.: The "3x + 1" problem and finite automata. Bull. EATCS **46**, 182–185 (1992)
21. Stérin, T., Woods, D.: The Collatz process embeds a base conversion algorithm. In: Schmitz, S., Potapov, I. (eds.) 14th International Conference on Reachability Problems. LNCS, Springer (2020, to appear). https://arxiv.org/abs/2007.06979v2
22. Stérin, T.: Binary expression of ancestors in the Collatz graph (2020). Expanded version. https://arxiv.org/abs/1907.00775v4
23. Terras, R.: A stopping time problem on the positive integers. Acta Arithmetica **30**(3), 241–252 (1976). http://eudml.org/doc/205476
24. Wirsching, G.J.: The Dynamical System Generated by the 3n + 1 Function. Springer, Heidelberg (1998). https://doi.org/10.1007/BFb0095985

The Collatz Process Embeds a Base Conversion Algorithm

Tristan Stérin[(✉)] and Damien Woods[(✉)]

Hamilton Institute, Department of Computer Science,
Maynooth University, Maynooth, Ireland
{tristan.sterin,damien.woods}@mu.ie
https://dna.hamilton.ie

Abstract. The Collatz process is defined on natural numbers by iterating the map $T(x) = T_0(x) = x/2$ when $x \in \mathbb{N}$ is even and $T(x) = T_1(x) = (3x + 1)/2$ when x is odd. In an effort to understand its dynamics, and since Generalised Collatz Maps are known to simulate Turing Machines [Conway, 1972], it seems natural to ask what kinds of algorithmic behaviours it embeds. We define a quasi-cellular automaton that exactly simulates the Collatz process on the square grid: on input $x \in \mathbb{N}$, written horizontally in base 2, successive rows give the Collatz sequence of x in base 2. We show that vertical columns simultaneously iterate the map in base 3. This leads to our main result: the Collatz process embeds an algorithm that converts any natural number from base 3 to base 2. We also find that the evolution of our automaton computes the parity of the number of 1s in any ternary input. It follows that predicting about half of the bits of the iterates $T^i(x)$, for $i = O(\log x)$, is in the complexity class NC^1 but outside AC^0. These results show that the Collatz process is capable of some simple, but non-trivial, computation in bases 2 and 3, suggesting an algorithmic approach to thinking about prediction and existence of cycles in the Collatz process.

Keywords: Collatz map · Model of computation · Reachability problem

1 Introduction

The Collatz process is defined on natural numbers by iterating the map $T(x) = T_0(x) = x/2$ when $x \in \mathbb{N}$ is even and $T(x) = T_1(x) = (3x + 1)/2$ when x is odd. The Collatz conjecture states that for all $x \in \mathbb{N}$ a finite number of iterations lead to 1. We know that 1-variable generalised Collatz maps (iterated linear equations of a single natural number variable with arbitrary mod) are capable of running arbitrarily algorithms [8], modulo an exponential simulation

Research supported by the European Research Council (ERC) under the EU's Horizon 2020 research & innovation programme, grant agreement No 772766, Active-DNA project, and Science Foundation Ireland (SFI), grant number 18/ERCS/5746.

S. Schmitz and I. Potapov (Eds.): RP 2020, LNCS 12448, pp. 131–147, 2020.
https://doi.org/10.1007/978-3-030-61739-4_9

scaling, which motivates our point of view in this paper: how complicated are the dynamics of the Collatz process? Perhaps the reason this process has resisted understanding is that it embeds algorithm(s) that solve problems with high computational complexity? Or perhaps by showing that this is not the case, we can get a handle on its dynamics?

We study the structure of iterations of the Collatz process both in binary and in ternary. We define a 2D quasi-cellular automaton (CQCA) that consists of a local rule, and a non-local[1] rule. A natural number is encoded as a binary string input to the CQCA, whose subsequent dynamics exactly execute the Collatz process with one horizontal row per odd iterate. Simultaneously, vertical columns simulate all iterations (both odd and even), but in base 3.

Results. Our main result, Theorem 11, is that the CQCA embeds a base conversion algorithm that can convert any natural number x from base 3 to base 2. This base conversion algorithm is natural and efficient, running in $\Theta(\log x)$ Collatz iterations. The result puts strict constraints on the short-term dynamics of the Collatz process and enables us to characterize the complexity of natural prediction problems on the dynamics of the CQCA: predicting about half of the bits of the iterates $T^i(x)$, for $i = O(\log x)$, is in the complexity class NC^1 but outside AC^0 (see Fig. 4). Our algorithmic perspective is suggestive of an open problem: *small Collatz cycles*: Does there exist $x > 2 \in \mathbb{N}$ such that $x = T^{\leq \lceil \log_2 x \rceil}(x)$?

The proofs of our main result and supporting lemmas use the fact that the *local* (CA-like) component of the CQCA can be thought of as simultaneously iterating two finite state transducers (FSTs). One FST computes $x/2$ in ternary on vertical columns, the other FST computes $3x+1$ in binary on horizontal rows, as shown in Fig. 3. Interestingly, the two FSTs are dual in the sense that states of one are symbols of the other, and arrows of one are read/write instructions of the other. In addition, intuitively, the *non-local* component of the CQCA initiates, or "bootstraps", these two FSTs by providing the location of the least significant bit to each.

Related and Future Work. Since the operations $3x + 1$ and $x/2$ have natural base 2 interpretations, studying the process in binary has been fruitful [4,7,11,24]. For example, in binary, predecessors in the Collatz process are characterised by regular expressions [25]: for each $x, k \in \mathbb{N}$ there is a regular expression, of size exponential in k, that characterises the binary representations of all $y \in \mathbb{N}$ that lead to x via k applications of T_1 and any number of T_0.

Cloney, Goles and Vichniac give a unidimensional "quasi" CA that simulates the Collatz process [6]. Their system works in base 2, for any $x \in \mathbb{N}$ given as an input in base 2, successive downward rows encode the iterates $x, T(x), T^2(x), \ldots$, in binary. The choice of whether to apply $x/2$ and $3x + 1$ is done explicitly based on the least significant bit, hence the rule is not local. In a similar spirit,

---

[1] The CQCA could easily be altered to remove the non-local rule and have it be a CA, but the obvious ways to do so would involve using more states and make the mapping to the Collatz process less direct. The CQCA has the freezing property [9,27]; states change obey a partial order, with a constant number of changes permitted per cell.

Bruschi [3] defines two distinct non-local 1D CA-like rules, one which works in base 2 and the other in base 3. Finally, in base 6, the Collatz process can be expressed as a local CA because carry propagation does not occur [13,15]. However, because these systems do not include carries in the automaton state's space, our base conversion result, and parity-based observations on the structure of Collatz iterates, are not apparent, nor is the CQCA-style embedding of dual FSTs that compute $3x + 1$ and $x/2$.

Base conversion is a problem which has been studied in several ways: it is known to be computable by iterating Finite State Transducers [2], it is known to be in the circuit complexity class NC^1 [10] and the complexity of computing base conversion of real numbers (infinite expansion) has been explored [1,2,12]. While we know that our framework generalises well to the Collatz process running on infinite binary strings (i.e. the extension [17,19,23] of T to $\mathbb{Z}_2$, see Remark 9), we leave to future work to show how our base conversion result applies in that case: for instance, can the CQCA convert any element $x \in \mathbb{Z}_3 \cap \mathbb{Z}_2 \cap \mathbb{Q}$ from $\mathbb{Z}_3$ (i.e. infinite ternary expansion) to $\mathbb{Z}_2$ (i.e. infinite binary expansion)?

Generalised Collatz maps, and related iterated dynamical systems, of both one and two variables, simulate Turing machines [8,14,16,20–22]. With two variables these maps simulate Turing machines in real time, just one map application per Turing machine step, and so have a P-complete prediction problem. The situation is less clear with 1-variable generalised Collatz maps (1D-GCMs), of which Collatz is an instance. Conway [8] showed that 1D-GCMs simulate Turing machines, but via an exponential slowdown, and it remains as a frustrating open problem whether the simulation can be made to run in polynomial time.[2] As with the 2-variable case, Turing Machines simulate 1D-GCMs in polynomial-in-n time (giving an upper bound), but a matching lower bound for predicting n steps of a 1D-GCM on an n-digit input remains elusive. Is the problem in NC, or even in NL? Is it P-complete? This point of view provides the motivation for a quest to understand the kinds of algorithmic behaviour that might be embedded in 1D-GCMs, and in their prime example, the Collatz map. Here, we show an AC^0 lower bound on that prediction problem in the CQCA model, and an NC^1 upper bound on roughly half the bits produced over $O(n)$ iterations. Finding a separation between the Collatz process itself, and 1D-GCMs, is an obvious next step. The structure we find in Collatz iterations leads us to guess that the CQCA n-step prediction problem might be simpler than the analogous problem for 1D-GCMs. We leave that challenging problem for future work.

We contend that our approach gives a fresh perspective that could lead to progress in understanding the Collatz process. We illustrate with two examples. Firstly, the CQCA runs in a maximally parallel fashion along a 135° diagonal (see Fig. 2(b)), thus our main base conversion result (Theorem 11) implies that such a diagonal encodes a natural number being converted from base 3 to 2. Hence, it implies that the Collatz process can be interpreted as running along successive

---

[2] The problem is closely related to the 2-counter machine problem: when simulating Turing machines, are 2-counter machines exponentially slower than 3-counter machines? See, for example, [14,18,22].

CQCA *diagonals* (rather than rows/columns), giving a new perspective on its dynamics, that we leave to future work. Secondly, by Theorem 23, and illustrated in Fig. 4, if we pick any cell along a column, the entire upper rectangle (above and to the left of the cell) is tightly characterised by our results: it is a base 3-to-2 conversion diagram (computable in NC^1, but not in AC^0). This leaves as future work the lower rectangle (below and to the left) and, we believe, embeds the full complexity of computing n forward Collatz steps and answering the *small Collatz cycles* conjecture. Section 4, Fig. 4 and Fig. 5 contain some additional discussion.

A simulator, `simcqca`, was built in order to run the CQCA[3]. The reader is invited to experience the results of this paper through the simulator.

2   The Collatz 2D Quasi-Cellular Automaton

The Collatz Process. Let $\mathbb{N} = \{0, 1, \dots\}$, $\mathbb{N}^+ = \{1, 2, \dots\}$ and let $\mathbb{Z}$ be the integers. The Collatz process consists of iterating the Collatz map $T : \mathbb{N} \to \mathbb{N}$ where $T(x) = T_0(x) = x/2$ if x is even and $T(x) = T_1(x) = (3x + 1)/2$ if x is odd. The Collatz conjecture states that for any $x \in \mathbb{N}$ the process will reach 1 after a finite number of iterations. The cyclic Collatz conjecture states that the only strictly positive cycle is $(1, 2, 1, \dots)$.

CQCA Definition. The CQCA is pictorially defined in Fig. 1 and more formally defined in Sect. 2.3 of the full version of this paper [26]. A *configuration* is defined on $\mathbb{Z}^2$, where each cell in $\mathbb{Z}^2$ has a state $(s, c) \in S = \{0, 1, \perp\}^2 \setminus \{(\perp, 0), (\perp, 1)\}$ containing *sum bit* s and *carry bit* c, said to be *defined* if 0/1 or *undefined* otherwise ($\perp$). We say that the cell is *defined* if both the sum and carry defined, *half-defined* if sum is defined and carry undefined, and *undefined* if sum and carry are undefined. Note, in all Figures, cell colour distinguishes between a carry bit being undefined or being 0, and empty light/dark grey cells have sum bit 0, as defined in Fig. 1(a).

A configuration update step is parallel and synchronous: First, the non-local rule is applied (at most 1 row per step is updated by the non-local rule) then, on the updated configuration, the local rule is applied everywhere it can be. The non-local rule implements the $+1$ part of the $3x + 1$ operation as follows: on any horizontal row which has only half-defined cells, with exactly one cell ρ having sum bit 1, then the neighbour immediately to the right of ρ gets, *ex nihilo*, a carry bit equal to 1 (shown in red in Fig. 1(a), see [26] for a formal definition). The local rule works as shown in Fig. 1(a): for any undefined cell $e \in \mathbb{Z}^2$, with a half-defined north neighbour with sum s_1, and defined north-east neighbour with sum s_0 and carry c_0, e's sum bit becomes $s_2 := s_0 + c_0 + s_1$ mod 2. Simultaneously, the carry, $c_1 := (s_0 + c_0 + s_1 \geq 2)$, is placed on the cell to the north of e. Figure 2(b) shows that the "natural" evolution frontier of the CQCA is along a 135° diagonal. Let the unit cardinal vectors in $\mathbb{Z}^2$ be $\texttt{EAST} = (1, 0)$, $\texttt{WEST} = (-1, 0)$, $\texttt{NORTH} = (0, 1)$, $\texttt{SOUTH} = (0, -1)$.

---

[3] Comprehensive instructions and tutorial at: https://github.com/tcosmo/simcqca.

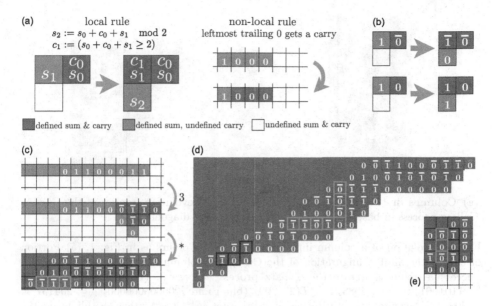

Fig. 1. CQCA definition and examples. (a) CQCA local rule, and non-local rule, for sum bit (s_2: written as 0 or 1), and carry bit (c_1: over-bar for $c_1 = 1$, omitted for $c_1 = 0$), where the bits c_0, s_0 and s_1 are already defined. Cell colour distinguishes between carry bit being undefined or 0, and empty light/dark grey cells have sum bit 0. (b) Two examples of the local rule. (c) Example run of the CQCA on base 2 (horizontal) input $w = 1100011$, showing the initial configuration $c_0[w]$ (Definition 2), then 3 steps, then > 20 steps. The CQCA evolves in the southwest direction, with successive rows corresponding to odd Collatz iterates in binary (Lemma 10), in this case (**99, 149**, 224, 112, 56, 28, 14, **7**, . . .), i.e., $[\![1100011]\!]_2 = 99$ (magenta), $[\![10010101]\!]_2 = 149$ (blue) and $[\![111]\!]_2 = 7$. (d) Part of the limit configuration $c_\infty[w]$ (Lemma 4) associated to $w = 1100011$. The trivial cycle (1,2), in blue, has been reached. In orange, an instance of the base conversion theorem (Theorem 11): north of the green cell (*anchor cell*) reads in base 3' (Definition 1), $[\![\bar{0}\bar{0}\bar{0}]\!]_{3'} = [\![111]\!]_3 = 13$ and is equal to the base 2 number represented by the sum bits to the west of the green cell: $[\![1101]\!]_2 = 13$. (e) Example run of the CQCA on vertical base 3 input $\alpha = 111$ encoded in base 3' as $\bar{0}\bar{0}\bar{0}$ (Definition 1). Each column is a successive T-iterate, from magenta to orange, we read: $[\![\bar{0}\bar{0}\bar{0}]\!]_{3'} = 13$, $[\![\bar{1}0\bar{1}]\!]_{3'} = 20 = T(13)$, $[\![\bar{0}\bar{0}\bar{0}]\!]_{3'} = 10 = T(20)$ (Lemma 10), see Fig. 2(a) for a larger vertical example. (Color figure online)

Base 2, 3 and 3'. Let $\{0,1\}^*$ be the set of finite binary strings, $\{0,1,2\}^*$ be the set of finite ternary strings. We index strings from their rightmost symbol, and $|\cdot|$ denotes string length, meaning we write, for instance, $w = w_{|w|-1} \ldots w_1 w_0 \in \{0,1\}^*$. We use the standard interpretation of strings from $\{0,1\}^*$ and $\{0,1,2\}^*$ as, respectively, base 2 and base 3 encodings of natural numbers where the rightmost symbol is the least significant digit. We write $[\![\cdot]\!]_2 : \{0,1\}^* \to \mathbb{N}$ and $[\![\cdot]\!]_3 : \{0,1,2\}^* \to \mathbb{N}$ for those interpretations. For instance, $[\![110]\!]_2 = [\![20]\!]_3 = 6$.

The CQCA uses a particular encoding for base 3 strings, called base 3', over the four-symbol alphabet $\{0, \bar{0}, 1, \bar{1}\}$. The CQCA states $(0,0)$, $(0,1)$, $(1,0)$,

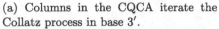

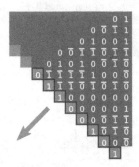

(a) Columns in the CQCA iterate the Collatz process in base 3′.

(b) The evolution frontier of the CQCA is a 135° diagonal.

Fig. 2. Evolution of a column input in the CQCA. Colours as in Fig. 1. (a) Portion of $c_\infty[\alpha]$, the limit configuration of the CQCA on column input $\alpha = 111211101211$. Successive columns iterate the Collatz process in base 3′ (Lemma 10). We read: $[\![\bar{1}0\bar{1}\bar{1}0\bar{1}00\bar{1}\bar{1}0\bar{1}]\!]_{3'} = 408314 = T(272209)$ (blue) and $[\![\bar{0}00\bar{0}0\bar{0}00\bar{0}0\bar{0}]\!]_{3'} = 204157 = T(408314)$ (orange). (b) Evolution of $c_0[\alpha]$ after $|\alpha|$ CQCA steps, highlighting the "natural" frontier of evolution of the CQCA as a 135° diagonal (blue cells). Cells to the north-east of the diagonal are defined, cells to the south-west are undefined and cells on the diagonal are half-defined. (Color figure online)

$(1,1) \in S$ respectively represent $0, \bar{0}, 1, \bar{1}$, using a (sum-bit, carry-bit) notation. The function $[\![\cdot]\!]_{3'\to3} : \{0,\bar{0},1,\bar{1}\}^* \to \{0,1,2\}^*$ converts base 3′ to base 3 in a straightforward symbol-by-symbol fashion: $0 \mapsto 0$, $\bar{0} \mapsto 1$, $1 \mapsto 1$ and $\bar{1} \mapsto 2$. For instance, $[\![\bar{0}\bar{0}01\bar{1}]\!]_{3'\to3} = 11012$. We write $[\![\cdot]\!]_{3'} = [\![[\![\cdot]\!]_{3'\to3}]\!]_3$ as the interpretation of base 3′ strings as natural numbers. However, because there are two choices for encoding the trit 1 in base 3′, converting from base 3 to base 3′ requires a definition:

Definition 1 (Base 3 to 3′ encoding). The function $[\![\cdot]\!]_{3\to3'} : \{0,1,2\}^* \to \{0,\bar{0},1,\bar{1}\}^*$ encodes a base 3 word α as a base 3′ word as follows: 0 is encoded as 0, 2 is encoded as $\bar{1}$, and 1 is encoded as 1 when the rightmost neighbour different from 1 is a 2, and as $\bar{0}$ otherwise. E.g., $[\![111211101211]\!]_{3\to3'} = 111\bar{1}0\bar{0}001\bar{1}00$.

Transducers and Duality. The CQCA rule has a local and non-local component. The *local* component simulates two FSTs: the $3x + 1$ FST in binary along horizontal rows and the $x/2$ FST in ternary along vertical columns (Fig. 3). Intuitively, the *non-local* component of the CQCA provides the least significant bit to these FSTs, in other words, it runs $x/2$ in binary (removing a trailing 0) and $3x + 1$ in ternary (adding a trailing 1). Interestingly, these two FSTs are closely related, we say they are *dual*: states of one are symbols of the other and that arrows of one are read/write instructions of the other. The proof of our main result Theorem 11, and of Lemmas 7 and 10, use the ability of the CQCA to simulate these FSTs simultaneously, one horizontally and the other vertically.

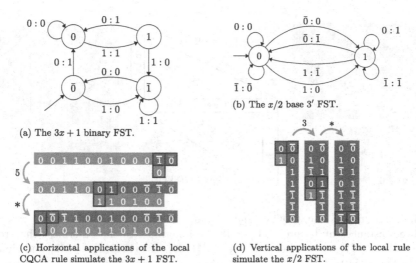

(a) The $3x + 1$ binary FST.

(b) The $x/2$ base $3'$ FST.

(c) Horizontal applications of the local CQCA rule simulate the $3x + 1$ FST.

(d) Vertical applications of the local rule simulate the $x/2$ FST.

Fig. 3. Panels (a) and (b) present the $3x + 1$ binary Finite State Transducer (FST) and the $x/2$ base $3'$ FST. Instructions "$s_1 : s_2$" mean "read s_1, write s_2". The FSTs are *dual* to one another: states of one are symbols of the other and arrows of one are read/write instructions of the other. Panels (c) and (d) show the relation with the local CQCA rule: (c) horizontal CQCA applications correspond to simulating the $3x+1$ FST and (d) vertical CQCA applications to simulating the $x/2$ FST. The same colour code as Fig. 1 is used. More precisely, in (c), on input $[\![00110010001]\!]_2 = 401$ and initial state $\bar{0}$ (rightmost 0 with red carry), we get output $[\![10010110100]\!]_2 = 1204 = 3 \times 401 + 1$. In (d) on input $[\![\bar{0}011\bar{1}\bar{1}0]\!]_{3'} = 862$ and initial state 0 (top-left most sum bit 0), we get output $[\![01\bar{1}0\bar{1}\bar{1}\bar{1}]\!]_{3'} = 431 = 862/2$. For clarity of exposition, we are "illegally" running the CQCA in a vertical (c), or horizontal (d), sequential mode—in the legal, or "natural", parallel mode, see Fig. 2(b), more cells would have been computed than are shown. The non-local component of the CQCA rule provides the FSTs with the (red) carry at the least significant digit. (Color figure online)

3   Base Conversion in the Collatz Process

Definition 2 (Binary initial configuration $c_0[w]$). For any binary input $w \in \{0, 1\}^*$, we define $c_0[w] \in S^{\mathbb{Z}^2}$ to be the initial configuration of the CQCA with w written on the horizontal ray $y = 0, x < 0$ as follows: for $1 \leq i \leq |w|$ we set $c_0[w](-i, 0) = (w_{i-1}, \perp)$, for $i > |w|$ we set $c_0[w](-i, 0) = (0, \perp)$ and for all other positions $(x, y) \in \mathbb{Z}^2$ we set $c_0[w](x, y) = (\perp, \perp)$.

Definition 3 (Ternary initial configuration $c_0[\alpha]$). For any ternary input $\alpha \in \{0, 1, 2\}^*$ we define $c_0[\alpha] \in S^{\mathbb{Z}^2}$ to be the initial configuration of the CQCA with $\alpha' = [\![\alpha]\!]_{3\to3'}$ (Definition 1) written on the vertical ray $x = 0, y > 0$ as follows: for $1 \leq i \leq |\alpha|$ we set $c_0[\alpha](0, i) = \text{state}(\alpha'_{i-1})$ where $\text{state} : \{0, \bar{0}, 1, \bar{1}\} \to S$ is such that $\text{state}(0) = (0,0)$, $\text{state}(\bar{0}) = (0,1)$, $\text{state}(1) = (1,0)$ and $\text{state}(\bar{1}) = (1,1)$. Also, for all $x < 0$ we set $c_0[\alpha](x, |\alpha|) = (0, \perp)$ and finally at all other positions we set $c_0[\alpha](x, y) = (\perp, \perp)$.

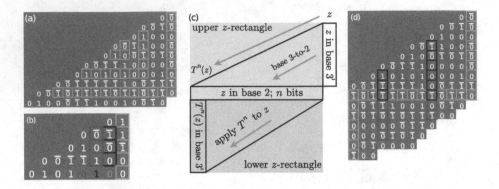

Fig. 4. Structure of the CQCA, and Collatz iterations, implied by our results. (a) Three instances of Theorem 11. From the innermost to the outermost they read $[\![\overline{0}]\!]_{3'} = [\![1]\!]_3 = [\![1]\!]_2 = 1$, then $[\![11\overline{1}0]\!]_{3'} = [\![1120]\!]_3 = [\![101010]\!]_2 = 42$, and finally $[\![\overline{0}0000\overline{0}0\overline{1}]\!]_{3'} = [\![101111011011]\!]_2 = 3035$. Anchor cells are in green. (b) Zoom-in of base conversion diagram of $[\![11\overline{1}0]\!]_{3'}$ (middle example in (a)). Each column, excluding its bottom cell, represents successive integer divisions by 2, e.g., from orange to magenta columns we read $[\![1120]\!]_3 = 42$, $[\![210]\!]_3 = 21$ and $[\![101]\!]_3 = 10$. At each step, the sum bits of the bottom row "store" the parity information of the previous column that was divided by 2: the orange bit gives the parity of the orange column and so on. Checking parity in base $3'$ is checking the parity of the number of 1s (both sum and carry bits), hence outside AC^0 (Corollary 15, Theorem 19). (c) For any input $z \in \mathbb{N}$ the schematic shows $n = \lceil \log_2 x \rceil$ iterations of the Collatz map T. The values $x, T(x), T^2(x), \ldots, T^n(x)$ appear as n columns, written in ternary (base $3'$). The configuration is cut by a horizontal line, whose cells encode z in binary (Theorem 11). The 'base conversion' upper z-rectangle is simple to define in terms of z, and is computable in NC^1. The lower z-rectangle embeds the full complexity of computing n iterations of T, but with input being in base 2 and output in base $3'$. (d) The influence of $z = [\![\overline{0}0000\overline{0}0\overline{1}]\!]_{3'} = [\![11011102]\!]_3 = 3035$ on its next $n = \lceil \log_2 z \rceil$ Collatz iterates. The column in orange on the right is the base $3'$ encoding of z. By Lemma 10, the column in orange on the left is the base $3'$ encoding of $T^n(z) = T^{12}(z) = [\![2020002]\!]_3 = 1622$. By Theorem 11, the orange horizontal sum bits give the base 2 representation of z. Hence the cells outlined in blue (upper z-rectangle) in $T^4(z)$ (second outlined column to the right) and $T^9(z)$ (third outlined column), are entirely determined by the base $3'$ to base 2 conversion diagram of z, only the cells outlined in magenta (lower z-rectangle) are independent from the base conversion algorithm. (Color figure online)

Each initial configuration has a well-defined unique limit configuration:

Lemma 4 (Limit configurations $c_\infty[w]$ and $c_\infty[\alpha]$). Let $w \in \{0,1\}^*$ be a finite binary string and $\alpha \in \{0,1,2\}^*$ be a finite ternary string. Then, in the CQCA evolution from the initial configurations $c_0[w] \in S^{\mathbb{Z}^2}$, and $c_0[\alpha] \in S^{\mathbb{Z}^2}$, both sum and carry bits in any cells are final: if set, they never change. Hence, limit configurations $c_\infty[w] = \lim_{i \to \infty} F^i(c_0[w])$ and $c_\infty[\alpha] = \lim_{i \to \infty} F^i(c_0[\alpha])$ exist.[4]

Proof. Both the local and non-local rules of the CQCA can only be applied either on undefined cells or half-defined cells. Furthermore, if applied on an undefined cell, the cell becomes half-defined or defined, and when applied on a half-defined cell it becomes defined. Hence, the following partial order $(\perp, \perp) < (s, \perp) < (s, c)$ with $s, c \in \{0, 1\}$ holds on the states of the CQCA: it has the freezing property [9,27] and cells can change state at most twice. Limit configurations are well-defined by taking the final state of each cell. □

Example 5. Figure 1(c) (top) and (d) respectively show $c_0[w]$ and a portion of $c_\infty[w]$ for $w = 1100011$. Figure 2(a) shows a portion of $c_\infty[\alpha]$ for $\alpha = 111211101211$; by Definition 1 we have $[\![111211101211]\!]_{3 \to 3'} = 111\bar{1}0\bar{0}001\bar{1}0\bar{0}$.

Next, we define how to read base 2 strings on rows and base $3'$ on columns of the CQCA:

Definition 6 (Mapping rows and columns to natural numbers). Let $x_0, y_0 \leq 0$ and $c \in S^{\mathbb{Z}^2}$ be a configuration. A finite segment of defined cells along a horizontal row x_0 of c is said to *give* word $w \in \{0, 1\}^*$ if w is exactly the concatenation of the sum bits in these cells (LSB on right). An infinite horizontal row y_0 of c is said to *give* $w \in \{0, 1\}^*$ if there is a $k \geq 0$ such that the defined sum bits on y_0 are $0^\infty w 0^k$.

A contiguous segment of defined cells on the column of c with x-coordinate x_0 is said to *give* the base $3'$ word $q \in \{0, \bar{0}, 1, \bar{1}\}^*$ if q is exactly the concatenation of the base $3'$ symbols corresponding to each cell's state[5] (where the southmost cell gives the least significant trit).

Intuitively, the following lemma says that rows of CQCA encode odd Collatz iterates in binary. There is one odd iterate per row, Fig. 1(c).

Lemma 7 (Rows simulate Collatz in base 2). Let $z \in \mathbb{N}^+$ and let $w \in \{0, 1\}^* \setminus \{0\}^*$ be the standard base 2 representation of z and, by Lemma 4, let $c_\infty[w] \in S^{\mathbb{Z}^2}$ be the CQCA limit configuration on input w. Then, the horizontal row $y_0 \leq 0$ of $c_\infty[w]$ gives the base 2 representation of the $|y_0|^{\text{th}}$ odd term in the Collatz sequence of z (Definition 6).

Proof. Note that it is enough to show that row $y = -1$ of $c_\infty[w]$ gives the base 2 representation of the second odd term in the Collatz sequence of z and then inductively apply the argument to all rows $y < -1$ to get the result. We have $w \neq 0^n$, hence there is at least one 1 in w and so, on row $y = 0$ of

---

[4] Furthermore, although the following fact is not used in the rest of this paper, one can prove that limit configurations contain no half-defined cells: each cell is either defined or undefined.

[5] The CQCA defined states are $(0, 0), (0, 1), (1, 0), (1, 1)$ and they respectively map to base $3'$ symbols $0, \bar{0}, 1, \bar{1}$.

$c_\infty[w]$, the non-local rule of the CQCA (Fig. 1(a)) was applied exactly once, at position $x_0 = \max(\{x < 0 \mid c_\infty[w](x,0) = (s,c) \text{ with } s = 1\}) + 1$ and we have $c_\infty[w](x_0,0) = (0,1)$. Sum bits on row $y = 0$ up to column $x_0 - 1$ give the representation of $z' \in \mathbb{N}$ (Definition 6), the first odd term in the Collatz sequence of $z = [\![w]\!]_2$.

From position $x = x_0 - 1$ down to $x_0 = -\infty$, the local rule of the CQCA was applied to produce carries on row $y = 0$ and sum bits on row $y = -1$. Given the definition of the CQCA local rule, that computation corresponds exactly to applying the binary $3x+1$ FST (Fig. 3(a) and (c)): input is read in the sum bits of row $y = 0$ from $x = x_0 - 1$ down to $x = -\infty$, output is produced in the sum bits of row $y = -1$, the initial state is $\bar{0}$, corresponding to $c_\infty[w](x_0,0) = (0,1)$. Hence, by Lemma 5 in the full version of this paper [26], which asserts the correctness of the $3x+1$ FST computation, sum bits of row $y = -1$ from $x = -\infty$ to $x = x_0 - 1$ give the binary representation of $3z' + 1$; starting with a $(0)^\infty$ prefix and with LSB to the east. By ignoring the $m \geq 1$ trailing zeros on row $y = -1$ (there is at least one because $3z' + 1$ is even), we get the representation of $(3z' + 1)/2^m$, the second odd iterate in the Collatz sequence of z. Note the non $(0)^\infty$ part of row $y = -1$ is produced in $\leq |w| + 2$ steps in the CQCA.   □

Remark 8. Although Lemma 7 only deals with odd Collatz iterations, even Collatz iterations also naturally appear in the CQCA: even terms occurring between the ith and the $(i+1)^{\text{th}}$ odd Collatz iterates correspond to the trailing 0s on row $y = -(i+1)$ of the CQCA. For example, on the third row in Fig. 1(c) (bottom) we read $7 = [\![111]\!]_2$ but also, in the trailing 0s, all $2^n \cdot 7$ for $1 \leq n \leq 6$.

Remark 9. The Collatz process can be generalised to an uncountable class of numbers that includes both $\mathbb{N}$ and $\mathbb{Z}$: $\mathbb{Z}_2$, the ring of 2-adic integers which syntactically corresponds to the set of infinite binary words [5,17]. Given that generalisation, the Collatz process can be run on more exotic numbers such as $-\frac{4}{23} \in \mathbb{Z}_2 \cap \mathbb{Q}$ and the Collatz conjecture generalises as follows: all rational 2-adic integers eventually reach a cycle[6] (known as Lagarias' Periodicity Conjecture [17]). When running the $3x+1$ FST on infinite binary inputs, one can show that it is computing the $3x+1$ function on 2-adic integers (see [26], Appendix B). Hence, Lemma 7 and the CQCA framework in general, can be generalised for working with infinite binary inputs and the Collatz process in $\mathbb{Z}_2$.

Intuitively, the following lemma says that columns of the CQCA encode all Collatz iterates, even and odd, in ternary, as in Figs. 1(e) and 2(a).

Lemma 10 (Columns simulate Collatz in base 3'). Let $z \in \mathbb{N}^+$ and let $\alpha \in \{0,1,2\}^* \setminus \{0\}^*$ be the standard base 3 representation of z, and, by Lemma 4, let $c_\infty[\alpha] \in S^{\mathbb{Z}^2}$ be the CQCA limit configuration on input α. Then, the vertical column $x_0 < 0$ in $c_\infty[\alpha]$ gives the base 3' representation of $T^{|x_0|}(z)$ (as the defined cells down to, and excluding, the southmost cell with sum bit 0, Definition 6).

---

[6] The set $\mathbb{Z}_2 \cap \mathbb{Q}$ exactly corresponds to irreducible fractions with odd denominator and parity is given by the parity of the numerator. For instance, the first Collatz steps of $-\frac{4}{23}$ are: $(-\frac{4}{23}, -\frac{2}{23}, -\frac{1}{23}, \frac{10}{23}, \frac{5}{23}, \dots)$. It reaches the cycle $(\frac{5}{23}, \frac{19}{23}, \frac{40}{23}, \frac{20}{23}, \frac{10}{23}, \frac{5}{23} \dots)$.

Proof. Note that it is enough to show that column $x_0 = -1$ of $c_\infty[\alpha]$ gives the base $3'$ representation of $T(z)$, with $z = [\![\alpha]\!]_3$, and then inductively apply the argument to all columns $x < -1$ to get the result. By construction of $c_0[\alpha]$ (Definition 3) and because all bits are final (Lemma 4) we have that the sum bit of $c_\infty[\alpha](-1, |\alpha|)$ is 0 (more generally, for all $x < 0$ we have the sum bit of $c_\infty[\alpha](x, |\alpha|)$ is 0). From there, the local CQCA rule (Fig. 1) is applied to each position $(-1, y)$ with $1 \leq y \leq |\alpha|$. This application of the rule corresponds exactly to running the base $3'$ $x/2$ FST (Fig. 3(b) and (d)) by reading the input on the base $3'$ symbols of cells of column $x = 0$ (both sum and carry bits of these cells are defined in $c_0[\alpha]$), writing the base $3'$ output on the cells of column $x = -1$ starting from initial FST state 0 (corresponding to the sum bit of $c_\infty[\alpha](-1, |\alpha|)$ being 0). In that interpretation, the sum bit output to the south of a cell by the local CQCA rule corresponds to the new FST state after reading the east base $3'$ symbol, hence the sum bit s of cell $(-1, 0)$ corresponds to the state of the $x/2$ FST after reading all base $3'$ symbols of $[\![\alpha]\!]_{3 \to 3'}$ (the least significant digit is at position $(0, 1)$). Two cases, with $z = [\![\alpha]\!]_3$:

1. If $z \equiv 0 \mod 2$, by Lemma 6 in the full version of this paper [26], we have that the final state of the $x/2$ FST after reading $[\![\alpha]\!]_{3 \to 3'}$ is 0 and that the output word $\alpha' \in \{0, \bar{0}, 1, \bar{1}\}$ is such that $[\![\alpha']\!]_{3'} = [\![\alpha]\!]_3/2 = z/2$. Hence, we deduce that column $x = -1$, from $y = |\alpha|$ down to $y = 1$, gives the base $3'$ representation of $T(z) = z/2$ which is what we wanted.

2. If $z \equiv 1 \mod 2$, by Lemma 6 of [26], we have that the final state of the $x/2$ FST after reading $[\![\alpha]\!]_{3 \to 3'}$ is 1 and that the output word $\gamma \in \{0, \bar{0}, 1, \bar{1}\}$, which is written on column $x = 1$, $y = |\alpha|$ down to $y = 1$, is such that $[\![\gamma]\!]_{3'} = ([\![\alpha]\!]_3 - 1)/2 = (z - 1)/2$. Furthermore, for $e = (0, 0) \in \mathbb{Z}^2$, the sum bit of $c_\infty[\alpha](e + \text{WEST})$, which corresponds to the final state of the $x/2$ FST, is $s = 1$ and $c_0[\alpha](e) = (\perp, \perp)$. Hence, the non-local rule will be applied at position e giving $c_0[\infty](e) = (0, 1) = (s_e, c_e)$. Then, at the following CQCA step, since $s_e + c_e + s = 0 + 1 + 1 \geq 2$ we get a carry bit of 1 at (e+WEST), i.e. $c_0[\infty](e + \text{WEST}) = (1, 1)$. It means that on column $x = -1$, with cell at position $(-1, 0) = e + \text{WEST}$ we add the base $3'$ symbol $\bar{1}$ on the least significant side of γ (word γ was output by the FST to the north of position $(-1, 0)$). Hence, column $x = -1$, from row $y = |\alpha|$ down to $y = 0$, interprets as: $3 \cdot [\![\gamma]\!]_{3'} + 2 = 3\frac{z-1}{2} + 2 = \frac{3z+1}{2} = T(z)$. Which is what we wanted.

Hence we get that column $x = -1$ gives the base $3'$ representation of $T(z)$. From the above points, it is immediate that the cell directly below the base $3'$ expression of $T(z)$ on column $x = -1$ has sum bit equal to 0 and that all cells below are undefined, giving the end of the Lemma statement. Note that it requires at most $|\alpha| + 2$ simulation steps for the CQCA to compute that representation (at most two extra cells to the south are used). ☐

We now prove our main result: the natural number written in base $3'$ on a column is converted to base 2 on the row directly south-west to it, see Fig. 4.

Theorem 11 (Base 3-to-2 conversion). *Let $\alpha \in \{0, 1, 2\}^*$ be a finite ternary string. By Lemma 4 let $c_\infty[\alpha] \in S^{\mathbb{Z}^2}$ be the CQCA limit configuration on input α.*

Then, in $c_\infty[\alpha]$, any position $e = (x_0, y_0) \in \mathbb{Z}^2$ such that both cells $e + \texttt{NORTH}$ and $e + \texttt{WEST}$ are defined, is called an *anchor cell* and has the *base conversion property*: there exists $z \in \mathbb{N}$ such that the defined cells on column x_0 strictly to the north of e give (Definition 6) the base $3'$ representation of z and the cells on row $y = y_0$ strictly to the west of e give the base 2 representation of z.

Proof. A direct consequence of the proof of Lemma 10 is that, in $c_\infty[\alpha]$ a defined sum bit $s \in \{0, 1\}$ at position (x_1, y_0) with $x_1 < 0$ and $y_0 < |\alpha|$ gives the state of the $x/2$ FST (Fig. 3(b) and (d)) immediately after it reads all base $3'$ symbols from row $y = |\alpha|$ to $y = y_0 + 1$ on column with x-coordinate $x_1 + 1$ of $c_\infty[\alpha]$. As s is the final FST state, by Lemma 6 [26], we get that s is 0 if that base $3'$ represented number was even; otherwise (if odd) s is 1.

We also know that the output of the FST, i.e. symbols strictly to the north of (x_1, y_0) represent $\lfloor x/2 \rfloor$. Hence, one base conversion step was performed: x mod 2 is written in the sum bit s at position (x_1, y_0) and $\lfloor x/2 \rfloor$ is computed to the north of it. By induction, all the other base conversion steps are also performed to the west of (x_1, y_0) and we get the result for the anchor cell (x_0, y_0) with $x_0 = x_1 + 1$. $\qquad\square$

Remark 12. Figure 4(a) presents several instances of Theorem 11. Note that position $(0, 0)$ is always an anchor cell in $c_0[\alpha]$ meaning that the CQCA converts $[\![\alpha]\!]_{3 \to 3'}$ from base $3'$ to base 2. Hence, the CQCA can effectively convert any base $3'$ input to base 2. Figure 4(b) presents the details of such a base conversion and shows that the CQCA base conversion algorithm is rather natural: at each step the parity of the input is computed, then the input is divided by 2. Also, the algorithm is efficient: for $x \in \mathbb{N}$, it can be shown that $\lceil \log_2(x) \rceil + \lceil \log_3(x) \rceil$ CQCA steps are sufficient to convert x from base 3 to base 2.

Remark 13. Theorem 11 also implies that limit configurations of row inputs are very similar to limit configurations of column inputs. Indeed, for $z \in \mathbb{N}$, with base 2 representation $w \in \{0, 1\}^*$ and base 3 representation $\alpha \in \{0, 1, 2\}^*$, we have: for all $x \leq 0$ and $y \leq 0$, $c_\infty[w](x, y) = c_\infty[\alpha](x, y)$. Thus, for all $x \leq -|w|$ and $y < 0$, columns of $c_\infty[w]$ iterate the Collatz process in ternary and the base conversion property holds for any anchor cell. Hence, the base conversion property naturally appears in the CQCA, for both row and column inputs.

Remark 14. Theorem 11 implies that cells with state $(0, 1)$ because of the non-local rule (red carries in Fig. 1 and 4) have an interesting interpretation column-wise: they implement the operation $3x + 1$ in ternary. Indeed, in base 3 the operation $3x + 1$ is trivial: it consists of appending 1 (represented here by $\bar{0}$ via Definition 1) to the base $3'$ representation of x. Thus, the CQCA "stacks" trivial ternary steps ($3x + 1$) on the same column, and trivial binary steps ($x/2$, i.e. a shift in binary) on the same row (Remark 8).

Corollary 15 (Parity checking in Collatz). Let $\alpha \in \{0, 1, 2\}^*$ and, by Lemma 4, let $c_\infty[\alpha] \in S^{\mathbb{Z}^2}$ be the limit configuration of the CQCA on input α (written in base $3'$ on column $x = 0$). For any anchor cell (Theorem 11) at position $e \in \mathbb{Z}^2$, let $s \in \{0, 1\}$ be the sum bit of the cell at $e + \texttt{WEST}$. Then, s is the parity of the number written in base $3'$ on the column at $e + \texttt{NORTH}$ and going to the north (Definition 6): this number is even iff $s = 0$ and odd iff $s = 1$.

Proof. Immediately implied by the proof of Theorem 11: the sum bit s at position $e + \texttt{WEST}$ gives the state in which the $x/2$ FST was after reading base $3'$ symbols on the column to the north of e. By Lemma 6 in the full version of this paper [26], the bit s is the parity of the number given by that column.   □

Example 16. Figure 4(b) outlines instances of Corollary 15.

3.1   Computational Complexity of CQCA Prediction

In this section we leverage our previous results to make statements about the computational complexity of predicting the CQCA.

Definition 17 (Bounded CQCA prediction problem). Given (a) any ternary input α, of length $n \in \mathbb{N}$ with resulting CQCA limit configuration $c_\infty[\alpha]$, and (b) any $(x, y) \in \mathbb{Z}^2$, where $\max(|x|, |y|) = O(n)$, what is the state $c_\infty[\alpha](x, y)$?

Remark 18. The version of this prediction problem, where the question is to predict the state $c_\infty[\alpha](x, y)$ for any $(x, y) \in \mathbb{Z}^2$, is at least as hard as the Collatz conjecture which in CQCA terms states that the $(1, 2, 1, 2, \ldots)$ "glider" will eventually occur, cf. blue outlined cells in Fig. 1(d).

It is straightforward to see that the bounded CQCA prediction problem is in the complexity class P, we can also give a lower bound in terms of AC^0 which is the class of problems solved by uniform[7] polynomial size, constant depth Boolean circuits with arbitrary gate fanin [22]:

Theorem 19. The bounded CQCA prediction problem is in P and not in AC^0.

Proof. To see that the problem is in P, simply encode the initial configuration as input to a two-tape Turing machine and, in $O(n^2)$ time, run the simulation forward until we have filled the plane up to distance n from the input.

To show the problem is outside AC^0, let $v \in \{0, 1\}^*$, and let $\alpha \in \{0, 1, 2\}^*$ be the base 3 word such that $\alpha = v$. Let $c_\infty[\alpha]$ be the limit configuration of the CQCA on (column) input α written in base $3'$ (as usual) on column $x = 0$. Since there are no 2-symbols in v, by Definition 1, the base $3'$ encoding of v maps 0 to 0 and 1 to $\bar{0}$, an encoding straightforward to represent in binary, and straightforward to compute in AC^0. Let b be the sum bit of $c_\infty[\alpha](-1, 0)$, i.e.

---

[7] Here, uniform has the meaning used in Boolean circuit complexity: that members of the circuit family for an infinite problem (set of words) are produced by a suitably simple algorithm [10]. The class AC^0 is of interest as it is "simple" enough to be strictly contained in P, that is, $AC^0 \subsetneq P$ [22]. This enables us to give lower bounds to the computational complexity, or inherent difficulty, of some problems, such as the bounded CQCA prediction problem.

$(-1, 0) + \texttt{NORTH} + \texttt{EAST}$ is the first symbol of α, hence (well) within the bound $n = O(|\alpha|)$ set by Definition 17.

Deciding whether a natural number is odd or even is equivalent to checking the parity of its number of 1s written in base 3. In base $3'$, this translates to checking the parity of the total number of sum and carry bits. By Corollary 15, b is the parity of the natural number represented by α, equivalently, b is the parity of the number of 1s in v. Hence the CQCA solves the problem $\texttt{PARITY} = \{v \in \{0, 1\}^* \mid v$ has an odd number of 1s$\}$ on the input v, with the result placed at distance $2 < |\alpha|$ from the input word α. Since $\texttt{PARITY}$ sits outside the complexity class AC^0 [22], the Bounded CQCA prediction problem is not in AC^0.   □

Remark 20. The proof of the previous theorem shows how to use the CQCA to solve $\texttt{PARITY}$. In fact, we can say something stronger: in any CQCA configuration, each sum bit with defined $\texttt{NORTH} + \texttt{EAST}$ neighbour solves an instance of the $\texttt{PARITY}$ problem. See Fig. 4.

Definition 21 (Upper z-rectangle prediction problem). Let $\alpha \in \{0, 1, 2\}^*$, $z = [\![\alpha]\!]_3 \in \mathbb{N}$ and let $n = \lceil \log_2 z \rceil \in \mathbb{N}$. Let $c[\alpha]$ be the associated CQCA initial configuration (Definition 3) and $R = \{(x, y) \mid -n \leq x \leq 0, 0 \leq y \leq |\alpha|\} \subsetneq \mathbb{Z}^2$. The upper z-rectangle prediction problem asks: What are the states, in the limit configuration $c_\infty[\alpha]$, of all cells with positions $(x, y) \in R$.

Example 22. Figure 4(c) gives a schematic representation of R, the upper z-rectangle. Figure 4(d) and Fig. 5 each give and instance of R, respectively for $z = [\![101111011011]\!]_2 = 3035$ and $z = [\![11110010]\!]_2 = 242$.

NC^1 is the class of problems solved by uniform polynomial size, logarithmic depth (in input length) Boolean circuits with gate fanin ≤ 2 [10]. The proof of the following theorem intuitively comes from the fact that predicting the entire upper z-rectangle amounts to running $\lceil \log_3 z \rceil$ base conversions in parallel (the proof is in the full version of this paper [26], Theorem 28).

Theorem 23. The upper z-rectangle prediction problem is in NC^1, and is not in AC^0.

Open Problem 1 (Lower z-rectangle prediction problem). What is the complexity of filling out the lower z-rectangle? I.e. the rectangle defined by $M = \{(x, y) \mid -n \leq x \leq 0, -n \leq y \leq 0\} \subsetneq \mathbb{Z}^2$ (same notation as Definition 21 and shown on Fig. 4(c)). We know it is in P, and that it is not in AC^0 (Theorem 19). Can we get matching lower and upper bounds?

These prediction problems are closely related to what we call *small positive Collatz cycles*. Indeed, if predicting M (Open Problem 1) was easy, one could hope to easily answer whether there exists $x > 2$ such that $x = T^{\leq \lceil \log_2(x) \rceil}(x)$:

Fig. 5. A small positive cycle, almost. The rightmost magenta column encodes $z = [\![\bar{1}\bar{1}\bar{1}\bar{1}\bar{1}]\!]_{3'} = 242$ in base $3'$. By Lemma 10, the leftmost magenta column is $T^{\lceil \log_2(x) \rceil}(x) = [\![\bar{1}\bar{1}1\bar{1}\bar{1}]\!]_{3'} = [\![22122]\!]_3 = 233$. The numbers 242 and 233 differ in only one trit (in blue): they almost generate a small positive Collatz cycle (Definition 24). Theorem 11 tells us that the region above the magenta row is characterised by base conversion, what about the region below?

Definition 24 (Small positive Collatz cycles). Let $x \in \mathbb{N}^+$. We say that x generates a small positive Collatz cycle if $x = T^m(x)$ with $0 < m \leq \lceil \log_2(x) \rceil$.

Open Problem 2. There are no small positive Collatz cycles besides $(2, 1, 2, 1, \dots)$.

Example 25. Figure 5 shows that answering the question about small positive Collatz cycle seems challenging. Indeed, the Figure illustrates that for $x = [\![\bar{1}\bar{1}\bar{1}\bar{1}\bar{1}]\!]_{3'} = [\![22222]\!]_3 = 242$, we have $T^{\lceil \log_2(x) \rceil}(x) = [\![22122]\!]_3 = 233$. The two numbers differ only in one trit, it is *almost* a small positive Collatz cycle.

Remark 26. One can also reformulate the small positive Collatz cycles problem by saying that, although the Collatz process is able to compute base $3'$-to-2 conversion, it can never compute base 2-to-$3'$ in $\lceil \log_2(x) \rceil$ CQCA steps (except for $x = 2$). Note that the CQCA would have to produce base 3 trits, from most to least significant trit which seems very hard to in the $\lceil \log_2(x) \rceil$ time constraint.

4 Discussion: Structural Implications on Collatz Sequences

Figure 4 summarises some strong consequences of our results on the structure of Collatz sequences. First, since columns are iterating the Collatz process in base 3 (Lemma 10), the base conversion theorem implies that any $z \in \mathbb{N}$ is giving some rather specific constraints on the next $\lceil \log_2 z \rceil$ iterations of the Collatz process. Specifically, the upper z-rectangle, Fig. 4(b), which corresponds to the diagram of the conversion of z from base $3'$ to base 2, is constraining, on average, half of the trits (base 3 digits) of any iteration $T^{\leq \lceil \log_2 z \rceil}(z)$, Fig. 4(d).

Second, Theorem 23 tells us that this upper z-rectangle is easy to predict, in the sense that the entire region can be computed in NC^1. The computational complexity of lower z-rectangle prediction remains open.

Third, Fig. 4(b) illustrates how the parity checking result of Corollary 15 places constraints on future iterates. A sum bit at *any* position $e \in \mathbb{Z}^2$ of a configuration $c_\infty[\alpha]$ is constrained to be the parity of the number of 1 s in the entire column (both sums and carries) whose base is at the north-east of e.

We should note that these phenomena are occurring everywhere, at each Collatz iterate. Hence, although patterns have been notoriously hard to identify in the Collatz process, our results give a new lens which reveals some detail.

Acknowledgement. Sincere thanks to Olivier Rozier for fruitful interactions, and to the anonymous reviewers for helpful comments.

References

1. Adamczewski, B., Bugeaud, Y.: On the complexity of algebraic numbers I. Expansions in integer bases. Ann. Math. **165**(2), 547–565 (2007). https://doi.org/10.4007/annals.2007.165.547
2. Adamczewski, B., Faverjon, C.: Mahler's method in several variables II: Applications to base change problems and finite automata, September 2018. https://arxiv.org/abs/1809.04826
3. Bruschi, M.: Two cellular automata for the 3x+1 map (2005). https://arxiv.org/abs/nlin/0502061
4. Capco, J.: Odd Collatz sequence and binary representations, March 2019. https://hal.archives-ouvertes.fr/hal-02062503
5. Caruso, X.: Computations with p-adic numbers. Journées Nationales de Calcul Formel. Les cours du CIRM (2018). https://hal.archives-ouvertes.fr/hal-01444183
6. Cloney, T., Goles, E., Vichniac, G.Y.: The $3x + 1$ problem: a quasi cellular automaton. Complex Syst. **1**(2), 349–360 (1987)
7. Cloney, T., Goles, E., Vichniac, G.Y.: The $3x + 1$ problem: a quasi cellular automaton. Complex Syst. **1**(2), 349–360 (1987)
8. Conway, J.: Unpredictable iterations. In: Number Theory Conference (1972)
9. Goles, E., Ollinger, N., Theyssier, G.: Introducing freezing cellular automata. In: Exploratory Papers of Cellular Automata and Discrete Complex Systems (AUTOMATA 2015), pp. 65–73 (2015)
10. Hesse, W., Allender, E., Barrington, D.A.M.: Uniform constant-depth threshold circuits for division and iterated multiplication. J. Comput. Syst. Sci. **65**(4), 695–716 (2002)
11. Hew, P.C.: Working in binary protects the repetends of $1/3^h$: comment on Colussi's 'The convergence classes of Collatz function'. Theor. Comput. Sci. **618**, 135–141 (2016). https://doi.org/10.1016/j.tcs.2015.12.033
12. Jakobsen, S.K., Simonsen, J.G.: Liouville numbers and the computational complexity of changing bases. In: Anselmo, M., Della Vedova, G., Manea, F., Pauly, A. (eds.) Beyond the Horizon of Computability. CiE 2020. Lecture Notes in Computer Science, vol. 12098, pp. 50–62. Springer, Cham (2020). https://doi.org/10.1007/978-3-030-51466-2_5
13. Kari, J.: Cellular automata, the Collatz conjecture and powers of 3/2. In: Yen, H.C., Ibarra, O.H. (eds.) Developments in Language Theory. DLT 2012. Lecture Notes in Computer Science, vol. 7410, pp. 40–49. Springer, Heidelberg (2012). https://doi.org/10.1007/978-3-642-31653-1_5

14. Koiran, P., Moore, C.: Closed-form analytic maps in one and two dimensions can simulate universal turing machines. Theoret. Comput. Sci. **210**(1), 217–223 (1999). https://doi.org/10.1016/s0304-3975(98)00117-0

15. Korec, I.: The $3x + 1$ problem, generalized pascal triangles and cellular automata. Mathematica Slovaca **42**(5), 547–563 (1992). http://eudml.org/doc/32424

16. Kurtz, S.A., Simon, J.: The undecidability of the generalized Collatz Problem. In: Cai, J.Y., Cooper, S.B., Zhu, H. (eds.) Theory and Applications of Models of Computation. TAMC 2007. Lecture Notes in Computer Science, vol. 4484, pp. 542–553. Springer, Heidelberg (2007). https://doi.org/10.1007/978-3-540-72504-6_49

17. Lagarias, J.C.: The $3x+1$ problem and its generalizations. Am. Math. Mon. **92**(1), 3–23 (1985). http://www.jstor.org/stable/2322189

18. Minsky, M.: Computation: Finite and Infinite Machines. Prentice Hall Inc., Englewood Cliffs (1967)

19. Monks, K.G., Yazinski, J.: The autoconjugacy of the 3x+1 function. Discrete Math. **275**(1–3), 219–236 (2004). https://doi.org/10.1016/s0012-365x(03)00125-0

20. Moore, C.: Unpredictability and undecidability in dynamical systems. Phys. Rev. Lett. **64**(20), 2354 (1990)

21. Moore, C.: Generalized shifts: unpredictability and undecidability in dynamical systems. Nonlinearity **4**(2), 199 (1991)

22. Moore, C., Mertens, S.: The Nature of Computation. Oxford University Press, Oxford (2011)

23. Rozier, O.: Parity sequences of the 3x+1 map on the 2-adic integers and Euclidean embedding. Integers **19**, A8 (2019)

24. Shallit, J., Wilson, D.A.: The "3x + 1" problem and finite automata. Bull. EATCS **46**, 182–185 (1992)

25. Stérin, T.: Binary expression of ancestors in the Collatz graph. In: Schmitz, S., Potapov, I. (eds.) 14th International Conference on Reachability Problems. LNCS, Springer (2020). (To appear). https://arxiv.org/abs/1907.00775v4

26. Stérin, T., Woods, D.: The Collatz process embeds a base conversion algorithm (2020). Expanded version. https://arxiv.org/abs/2007.06979v3

27. Vollmar, R.: On cellular automata with a finite number of state changes. In: Knödel, W., Schneider, H.J. (eds.) Parallel Processes and Related Automata/Parallele Prozesse und damit zusammenhängende Automaten. Computing Supplementum, vol. 3, pp. 181–191. Springer, Vienna (1981). https://doi.org/10.1007/978-3-7091-8596-4_13

The Complexity of the Label-Splitting-Problem for Flip-Flop-Nets

Ronny Tredup[(✉)]

Institut für Informatik, Theoretische Informatik, Universität Rostock,
Albert-Einstein-Straße 22, 18059 Rostock, Germany
ronny.tredup@uni-rostock.de

Abstract. Let τ be a type of nets. *Synthesis* consists in deciding whether a given labelled transition system (TS) A can be implemented by a net N of type τ. In case of a negative decision, it may be possible to convert A into an implementable TS A' by relabeling edges that previously had the same label differently: *Label-splitting* is the problem to decide for a TS A and a natural number κ whether there is an implementable TS B with at most κ labels, which is derived from A by splitting labels. In this paper, we show that label-splitting is NP-complete if τ corresponds to the type of flip-flop nets or some flip-flop net derivatives.

1 Introduction

The so-called synthesis problem for nets of type τ (τ-synthesis, for short) consists in deciding whether for a given transition system (TS, for short) A there is a net N of type τ (a τ-net, for short) that implements A. In case of a positive decision, N should be constructed.

τ-Synthesis is used to, for example, extract concurrency from sequential specifications like TS and languages [5] and has applications in, for example, process discovery [1], supervisory control [11] and the synthesis of speed independent circuits [9].

However, whether N exists depends crucially on the implementation we are striving at, that is, whether N shall be an (exact) *realization*, meaning that A and N's reachability graph are isomorphic, or a *language-simulation* or an *embedding*. Unfortunately, for all implementations, a solution does not always exist. In that case, *label-splitting* [4,7,8,16] may be an option. This approach may convert a non-implementable TS A into an implementable one A' by relabeling edges that previously had the same label differently. However, the new events produced by the label-splitting increase the complexity of the net derived, since each new copy will be transformed into a transition. Hence, it is desired to find a label-splitting that induces the minimal number of transitions in the sought net. This allows to consider τ-*label-splitting* as a decision problem with input A and $\kappa \in \mathbb{N}$; the question is whether there is a TS B that, firstly, is

© Springer Nature Switzerland AG 2020
S. Schmitz and I. Potapov (Eds.): RP 2020, LNCS 12448, pp. 148–163, 2020.
https://doi.org/10.1007/978-3-030-61739-4_10

derived from A by splitting labels and, secondly, has at most κ labels and, more-over, is implementable by a τ-net. Recently, in [17], it has been shown that τ-label-splitting aiming at embedding is NP-complete if τ equals the type of *Place/Transition*-nets. Moreover, in [20], we have shown that τ-label-splitting aiming at language-simulation or realization is also NP-complete for this type. Naturally, it raises the question whether the problem is of a different complexity provided the net-type sought or the implementation sought or both are different.

A whole class of net-types for which such investigations are certainly of inter-est is defined by the family of the so-called *Boolean types of nets* [4,6,12–15,18], since the respective nets are widely accepted as excellent tools for modeling concurrent and distributed systems.

Boolean nets allow at most one token on each place p in every reachable mark-ing. Therefore, p is considered a Boolean condition that is *true* if p is marked and *false* otherwise. A place p and a transition t of a Boolean net N are related by one of the following Boolean *interactions*: *no operation* (nop), *input* (inp), *out-put* (out), *unconditionally set to true* (set), *unconditionally reset to false* (res), *inverting* (swap), *test if true* (used), and *test if false* (free). The relation between p and t determines which conditions p must satisfy to allow t's firing and which impact has the firing of t on p. Boolean nets are then classified by the inter-actions of $I = \{\text{nop}, \text{inp}, \text{out}, \text{res}, \text{set}, \text{swap}, \text{used}, \text{free}\}$ that they apply or spare. More exactly, a subset $\tau \subseteq I$ is called a *Boolean type of net* and a net N is of type τ (a τ-net) if it applies at most the interactions of τ.

However, the determination of the complexity of τ-label splitting does not actually define an open problem for all Boolean types of nets. In particular, it is known that τ-synthesis aiming at language-simulation or realization is NP-complete for 84 of the 128 thinkable Boolean types of nets that allow indepen-dence between places and transitions [3,19,22,23], that is, which contain nop. For these types, the NP-completeness of τ-synthesis implies already the NP-completeness of the corresponding label splitting problem by a trivial reduction from the former to the latter, which defines κ in a way that forbids any splitting. This puts the Boolean types with a tractable synthesis problem in focus. One of the most prominent Boolean type that fulfills this criterion is the so-called type of *flip-flop* nets $\tau = \{\text{nop}, \text{inp}, \text{out}, \text{used}\}$. Flip-flop nets have been originally introduced in [18] as the Boolean counterpart of the Place/Transition nets. The latter characterization is mainly based on the fact that synthesis aiming at flip-flop nets is solvable by a polynomial time algorithm [18] that is derived from the algorithm for synthesis aiming at Place/Transition nets, which has been intro-duced in [2]. Moreover, the algorithm for flip-flop nets [18] is extendable to all types $\tau = \{\text{nop}, \text{swap}\} \cup \omega$ with $\omega \subseteq \{\text{inp}, \text{out}, \text{used}, \text{free}\}$ [22], which makes their synthesis problem also tractable.

In this paper, for all types $\tau = \{\text{nop}, \text{swap}\} \cup \omega$ with $\omega \subseteq \{\text{inp}, \text{out}, \text{used}, \text{free}\}$, we investigate the computational complexity of τ-label-splitting for all intro-duced implementations: embedding, language-simulation and realization. In par-ticular, we show that label-splitting aiming at embedding is NP-complete for all of these types. Moreover, label-splitting aiming at language-simulation or realization is NP-complete if $\omega \neq \emptyset$, else it is tractable.

We obtain our NP-completeness results by reductions from a variant of the well-known vertex cover problem. Our current approach generalizes our methods from [20] and tailors them to flip-flop nets and its aforementioned derivatives. In contrast to our previous approach [20], for any fixed type, we get al.ong with one reduction, regardless of which of the implementations embedding, language-simulation or realization we aim for.

This paper is organized as follows. The next Sect. 2 introduces necessary notions and definitions. After that, Sect. 3 presents our complexity results. Finally, Sect. 4 briefly closes the paper. Due to space restrictions, one proof and some illustrations are moved to the appendix.

2   Preliminaries

This section introduces the basic notions used throughout the paper.

Transition Systems. A (deterministic, reduced) *transition system* (TS) $A = (S, \Sigma, \delta)$ is a directed labeled graph with the set of nodes S (called *states*), the set of labels Σ (called *events*) and partial *transition function* $\delta : S \times \Sigma \longrightarrow S$ such that for every $e \in \Sigma$ there are states $s, s' \in S$ such that $\delta(s, e) = s'$. For convenience, with a little abuse of notation, we often identify δ and the set $\{(s, e, s') \mid s, s' \in S, e \in \Sigma : \delta(s, e) = s'\}$. Event e *occurs* at s, denoted by $s \xrightarrow{e}$, if $\delta(s, e)$ is defined. We denote $\delta(s, e) = s'$ by $s \xrightarrow{e} s'$. An *initialized* TS $A = (S, \Sigma, \delta, \iota)$ is a TS with a distinct *initial* state $\iota \in S$ where every state $s \in S$ is *reachable* from ι by a directed labeled path. If $w = e_1 \ldots e_n \in \Sigma^*$, then by $\iota \xrightarrow{w}$ we denote that there are states $\iota = s_0, \ldots, s_n \in S(A)$ such that $s_i \xrightarrow{e_{i+1}} s_{i+1} \in A$ for all $i \in \{0, \ldots, n-1\}$. The *language* of A is defined by $L(A) = \{w \in \Sigma^+ \mid \iota \xrightarrow{w}\} \cup \{\varepsilon\}$. We refer to the components of a TS A by $S(A)$ (states), $\Sigma(A)$ (events), δ_A (transition function) and ι_A (initial state).

Simulations. Let A and B be TS with the same set of events Σ. We say B simulates A, if there is a mapping $\varphi : S(A) \to S(B)$ such that $s \xrightarrow{e} s' \in A$ implies $\varphi(s) \xrightarrow{e} \varphi(s') \in B$; such a mapping is called a *simulation* (between A and B). φ is an *embedding*, denoted by $A \hookrightarrow B$, if it is injective; φ is a *language-simulation*, denoted by $A \rhd B$, if $\varphi(s) \xrightarrow{e}$ implies $s \xrightarrow{e}$, implying $L(A) = L(B)$ [4, p. 67]; φ is an *isomorphism*, denoted by $A \cong B$, if it is a bijective language simulation.

Label-Splitting. Let $A = (S, \Sigma, \delta, \iota)$ be a TS and $e_1, \ldots, e_n \in \Sigma$ be pairwise distinct events. The *label-splitting* of the events $e_1, \ldots, e_n$ into the (pairwise distinct) events $e_1^1, \ldots, e_1^{m_1}, \ldots, e_n^1, \ldots, e_n^{m_n}$, where $m_j \geq 2$ for all $j \in \{1, \ldots, n\}$, yields the event set $\Sigma' = (\Sigma \setminus \{e_1, \ldots, e_n\}) \bigcup_{i=1}^n \{e_i^j \mid j \in \{1, \ldots, m_j\}\}$. A TS $B = (S, \Sigma', \delta', \iota)$ is a Σ'-*label-splitting* (Σ'-LS, for short) of A if $|\delta| = |\delta'|$ and, for all $s, s' \in S$ and all $e \in \Sigma$, the following is true: If $\delta(s, e) = s'$ and $e \notin \{e_1, \ldots, e_n\}$, then $\delta'(s, e) = s'$; if $\delta(s, e) = s'$ and $e = e_i$ for some $i \in \{1, \ldots, n\}$, then there is exactly one $\ell \in \{1, \ldots, m_i\}$ such that $\delta'(s, e_i^\ell) = s'$. We say that $\{e_1, \ldots, e_n\}$ is *the set of events of A that occur split in B*.

x	nop(x)	inp(x)	out(x)	set(x)	res(x)	swap(x)	used(x)	free(x)
0	0		1	1	0	1		0
1	1	0		1	0	0	1	

Fig. 1. All interactions i of I. If a cell is empty, then i is undefined on the respective x.

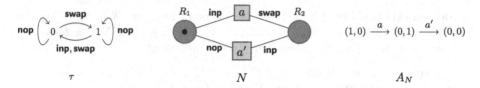

τ N A_N

Fig. 2. The type $\tau = \{\mathsf{nop}, \mathsf{inp}, \mathsf{swap}\}$ and a τ-net N and its reachability graph A_N.

$$t_0 \xrightarrow{a} t_1 \xrightarrow{a} t_2 \qquad\qquad t_0 \xrightarrow{a} t_1 \xrightarrow{a'} t_2 \qquad\qquad 1 \xrightarrow{\mathsf{inp}} 0 \xrightarrow{\mathsf{nop}} 0$$

$$A \qquad\qquad\qquad\qquad\qquad B \qquad\qquad\qquad\qquad\qquad B^{R_1}$$

Fig. 3. The TS A with event set $\Sigma = \{a\}$, a Σ'-label-splitting B of A, where $\Sigma' = (\Sigma \setminus \{a\}) \cup \{a, a'\}$ and the image B^{R_1} of the τ-region $R_1 = (sup_1, sig_1)$, where $sup_1(t_0) = 1$, $sup_1(t_1) = sup_1(t_2) = 0$, $sig_1(a) = \mathsf{inp}$ and $sig_1(a') = \mathsf{nop}$.

Boolean Types of Nets [4, pp. 139–152]. The following notion of Boolean types of nets allows to capture *all* Boolean nets in a *uniform* way. A *Boolean type of net* $\tau = (\{0, 1\}, \Sigma_\tau, \delta_\tau)$ is a TS such that Σ_τ is a subset of the *Boolean interactions*: $\Sigma_\tau \subseteq I = \{\mathsf{nop}, \mathsf{inp}, \mathsf{out}, \mathsf{set}, \mathsf{res}, \mathsf{swap}, \mathsf{used}, \mathsf{free}\}$. Each interaction $i \in I$ is a binary partial function $i : \{0, 1\} \to \{0, 1\}$ as defined in Fig. 1. For all $x \in \{0, 1\}$ and all $i \in \Sigma_\tau$, the transition function of τ is defined by $\delta_\tau(x, i) = i(x)$. Since a type τ is completely determined by Σ_τ, we often identify τ with Σ_τ.

τ-**Nets.** Let $\tau \subseteq I$. A Boolean net $N = (P, T, f, M_0)$ of type τ (a τ-*net*) is given by finite disjoint sets P of *places* and T of *transitions*, a (total) *flow function* $f : P \times T \to \tau$, and an *initial marking* $M_0 : P \longrightarrow \{0, 1\}$. A transition $t \in T$ can *fire* in a marking $M : P \longrightarrow \{0, 1\}$ if $\delta_\tau(M(p), f(p, t))$ is defined for all $p \in P$. By firing, t produces the marking $M' : P \longrightarrow \{0, 1\}$ where $M'(p) = \delta_\tau(M(p), f(p, t))$ for all $p \in P$, denoted by $M \xrightarrow{t} M'$. The behavior of τ-net N is captured by a transition system A_N, called the *reachability graph* of N. The states set $RS(N)$ of A_N consists of all markings that can be reached from initial state M_0 by sequences of transition firings.

Example 1. Figure 2 shows the type $\tau = \{\mathsf{nop}, \mathsf{inp}, \mathsf{swap}\}$ and the τ-net $N = (\{R_1, R_2\}, \{a, a'\}, f, M_0)$ with places R_1, R_2, flow-function $f(R_1, a) = f(R_2, a') = \mathsf{inp}$, $f(R_1, a') = \mathsf{nop}$, $f(R_2, a) = \mathsf{swap}$ and initial marking M_0 defined by $(M_0(R_1), M_0(R_2)) = (1, 0)$. Since $1 \xrightarrow{\mathsf{inp}} 0 \in \tau$ and $0 \xrightarrow{\mathsf{swap}} 1 \in \tau$, the transition a can fire in M_0, which leads to the marking $M = (M(R_1), M(R_2)) = (0, 1)$. After that, a' can fire, which results in the marking $M' = (M'(R_1), M'(R_2)) = (0, 0)$. The reachability graph A_N of N is depicted on the right hand side of Fig. 2.

Implementations. Let A be a TS, τ be a type of nets and N a τ-net. We say N is a (exact) *realization* of A if $A \cong A_N$. If $A \rhd A_N$, then N is a *language-simulation* of A; if $A \hookrightarrow A_N$, then N is an embedding of A.

τ-Regions. Let τ be a Boolean type of nets. If a TS A is implementable by a τ-net N, then we want to construct N purely from A. Since A_N has to simulate A, N's transitions correspond to A's events. The connection between global states in TS and local states in the sought net is given by *regions of TS* that mimic places: A τ-region $R = (sup, sig)$ of $A = (S, \Sigma, \delta, \iota)$ consists of the *support* $sup : S \to \{0, 1\}$ and the *signature* $sig : \Sigma \to \Sigma_\tau$ where every edge $s \xrightarrow{e} s'$ of A leads to an edge $sup(s) \xrightarrow{sig(e)} sup(s')$ of type τ. If $P = q_0 \xrightarrow{e_1} \dots \xrightarrow{e_n} q_n$ is a path in A, then $P^R = sup(q_0) \xrightarrow{sig(e_1)} \dots \xrightarrow{sig(e_n)} sup(q_n)$ is a path in τ. We say P^R is the *image* of P (under R). Notice that R is *implicitly* completely defined by $sup(\iota)$ and sig: Since A is reachable, for every state $s \in S(A)$, there is a path $\iota \xrightarrow{e_1} \dots \xrightarrow{e_n} s_n$ such that $s = s_n$. Thus, since τ is deterministic, we inductively obtain $sup(s_{i+1})$ by $sup(s_{i+1}) = \delta_\tau(sup(s_i), sig(e_{i+1}))$ for all $i \in \{0, \dots, n-1\}$ and $s_0 = \iota$. Consequently, we can compute sup and thus R purely from $sup(\iota)$ and sig, cf. Fig. 3. A region (sup, sig) models a place p and the associated part of the flow function f. In particular, $f(p, e) = sig(e)$ and $M(p) = sup(s)$, for marking $M \in RS(N)$ that corresponds to $s \in S(A)$. Every set $\mathcal{R}$ of τ-regions of A defines the *synthesized* τ-net $N_A^{\mathcal{R}} = (\mathcal{R}, \Sigma, f, M_0)$ with $f((sup, sig), e) = sig(e)$ and $M_0((sup, sig)) = sup(\iota)$ for all $(sup, sig) \in \mathcal{R}, e \in \Sigma$.

τ-State and τ-Event Separation. To ensure that the input behavior is captured by the synthesized net, we have to distinguish global states, and prevent the firings of transitions when their corresponding events are not present in TS. This is stated as so called *separation atoms* and *properties*. A pair (s, s') of distinct states of A defines a *states separation atom* (SSP atom). A τ-region $R = (sup, sig)$ *solves* (s, s') if $sup(s) \neq sup(s')$. If every SSP atom of A is τ-solvable then A has the τ-*states separation property* (τ-SSP, for short). A pair (e, s) of event $e \in \Sigma$ and state $s \in S$ where e does not occur, that is $\neg s \xrightarrow{e}$, defines an *event/state separation atom* (ESSP atom). A τ-region $R = (sup, sig)$

solves (e, s) if $sig(e)$ is not defined on $sup(s)$ in τ, that is, $\neg sup(s) \xrightarrow{sig(e)}$. If every ESSP atom of A is τ-solvable then A has the *τ-event state separation property* (τ-ESSP, for short). A set $\mathcal{R}$ of τ-regions of A is called a *τ-witness* for A's τ-SSP, respectively τ-ESSP, if for each SSP atom, respectively ESSP atom, there is a τ-region R in $\mathcal{R}$ that solves it. The next lemma ([4, p. 162], Proposition 5.10) establishes the connection between the existence of τ-witnesses and the existence of an implementing τ-net N in dependence of the testified property:

Lemma 1 ([4]). *If A is a TS, τ a Boolean type of nets and N a τ-net, then $A \hookrightarrow A_N$, respectively $A \rhd A_N$, respectively $A \cong A_N$ if and only if there is a τ-witness $\mathcal{R}$ that testifies A's τ-SSP, respectively τ-ESSP, respectively both τ-SSP and τ-ESSP, and $N = N_A^{\mathcal{R}}$.*

By Lemma 1, deciding the existence of an implementing net is equivalent to deciding if the input TS has the property that corresponds to the implementation. Our NP-completeness proofs substantially exploit this deep connection between the existence of a witness (for the demanded property) and the existence of an implementing system, which is testified by the witness. Finally, this leads to the following three decision problems that are the main subject of this paper:

LS-τ-EMBEDDING
Input: a TS $A = (S, \Sigma, \delta, \iota)$, a natural number κ.
Question: Does there exist a Σ'-LS B of A with $|\Sigma'| \leq \kappa$ that has the
            τ-SSP?

LS-τ-LANGUAGE SIMULATION
Input: a TS $A = (S, \Sigma, \delta, \iota)$, a natural number κ.
Question: Does there exist a Σ'-LS B of A with $|\Sigma'| \leq \kappa$ that has the
            τ-ESSP?

LS-τ-REALISATION
Input: a TS $A = (S, \Sigma, \delta, \iota)$, a natural number κ.
Question: Does there exist a Σ'-LS B of A with $|\Sigma'| \leq \kappa$ that has both the
            τ-SSP and the τ-ESSP?

Example 2. Let τ be defined like in Fig. 2 and A, B and B^{R_1} like in Fig. 3. The TS A has neither the τ-SSP nor the τ-ESSP, since the atoms (t_0, t_2) and (a, t_2) are not τ-solvable. The TS B is a Σ'-label-splitting of A, where $\Sigma' = (\Sigma \setminus \{a\}) \cup \{a, a'\}$, and has both the τ-SSP and ESSP. The region $R_1 = (sup_1, sig_1)$ that solves (t_0, t_1), (t_0, t_2), (a, t_1) and (a, t_2) is implicitly defined by $sup_1(t_0) = 1$, $sig_1(a) = \mathsf{inp}$ and $sig_1(a') = \mathsf{nop}$. We obtain sup_1 and thus R_1 explicitly by $sup_1(t_1) = \delta_\tau(1, \mathsf{inp}) = 0$ and $sup_1(t_2) = \delta_\tau(0, \mathsf{nop}) = 0$. B^{R_1} shows the image of B under R_1. The remaining (E)SSP atoms (t_1, t_2), (a', t_0) and (a', t_1) are solved by the following region $R_2 = (sup_2, sig_2)$ that is implicitly defined by $sup_2(t_0) = 0$, $sig_2(a) = \mathsf{swap}$ and $sig_2(a') = \mathsf{inp}$. The set $\mathcal{R} = \{R_1, R_2\}$ is a witness for the τ-(E)SSP of A and the net $N_A^{\mathcal{R}}$ is exactly the net N that is depicted in Fig. 2. N is a realization of A, since a bijective simulation φ between A and A_N is given by $\varphi(t_0) = (1,0)$, $\varphi(t_1) = (0,1)$ and $\varphi(t_2) = (0,0)$.

3   Our Contribution

The following theorem presents the main result of this paper:

Theorem 1. *If $\tau = \{nop, swap\} \cup \omega$ with $\omega \subseteq \{inp, out, used, free\}$, then (1)* LS-$\tau$-EMBEDDING *is NP-complete, and (2)* LS-τ-LANGUAGE SIMULATION *and* LS-τ-REALISATION *are NP-complete if $\omega \neq \emptyset$, otherwise they are in P.*

If $\tau = \{nop, swap\}$, then a TS A has the τ-ESSP if and only if every event occurs at every state. This is because both swap and nop are defined on both 0 and 1. Thus, any ESSP atom (e, s) of A would be unsolvable. This implies that a Σ'-label-splitting of A can never yield a TS that has the τ-ESSP, since it would produce unsolvable ESSP atoms. Moreover, if A is such that every event occur at every state, but has an unsolvable SSP atom, then label-splitting for this atom would destroy the ESSP. Hence, for this type, the decision problems are trivial, since either A is already implementable or it has to be rejected. Thus, for the proof of Theorem 1, it remains to consider the NP-completeness results.

The decision problems LS-τ-EMBEDDING, LS-τ-LANGUAGE SIMULATION and LS-τ-REALISATION are in NP: If a sought Σ'-label-splitting $B = (S, \Sigma', \delta', \iota)$ of a TS $A = (S, \Sigma, \delta, \iota)$ exists, then a Turing-machine M can compute B in a non-deterministic computation in time polynomial in the size $|\delta|$ of A, since $|\delta| = |\delta'|$. After that, M verifies in time polynomial in the size $|\delta'|$ of B (and thus of A) that it allows the sought implementation [18,22].

Our NP-completeness proofs base on reductions from the following classical variant of the vertex cover (VC) problem [10, p. 190]:

3-BOUNDED VERTEX COVER (3BVC)
Input: a Graph $G = (V, E)$ such every vertex $v \in V$ is a member of at most three distinct edges $e_0, e_1, e_2 \in E$, a natural number $\lambda \in \mathbb{N}$.
Question: Does there exist a λ-VC of G, that is, a subset $M \subseteq V$ with $|M| \leq \lambda$ and $M \cap e \neq \emptyset$ for all $e \in E$?

Example 3 (3BVC). The instance $(G, 2)$, where $G = (V, E)$ such that $V = \{v_0, v_1, v_2, v_3\}$ and $E = \{\{v_0, v_1\}, \{v_0, v_2\}, \{v_0, v_3\}, \{v_1, v_2\}, \{v_2, v_3\}\}$, is a yes-instance of 3BVC, since $M = \{v_0, v_2\}$ is a suitable vertex cover of G.

In the remainder of this paper, let (G, λ) be an input of 3BVC, where $G = (V, E)$ is a graph with n vertices $V = \{v_0, \ldots, v_{n-1}\}$ and m edges $E = \{e_0, \ldots, e_{m-1}\}$ such that $e_i = \{v_{i_0}, v_{i_1}\}$ and $i_0 < i_1$ for all $i \in \{0, \ldots, m-1\}$.

For the proof of Theorem 1, we reduce (G, λ) to a pair (A_τ, κ) of TS $A_\tau = (S, \Sigma, \delta, \perp_0)$ and natural number κ such that the following conditions are satisfied:

1. If there is a Σ'-label-splitting of A_τ that satisfies $|\Sigma'| \leq \kappa$ and has the τ-SSP or the τ-ESSP, then G has a λ-VC.

2. If G has a λ-VC, then there is a Σ'-label-splitting of A_τ that satisfies $|\Sigma'| \leq \kappa$ and has both the τ-SSP and the τ-ESSP.

Obviously, a polynomial-time reduction that satisfies Condition 1 and Condition 2 ensures that G allows a λ-VC if and only if A_τ allows a Σ'-label-splitting that satisfies $|\Sigma'| \leq \kappa$ and has the τ-SSP, the τ-ESSP or both, according to which property is sought. Hence, it proves Theorem 1.

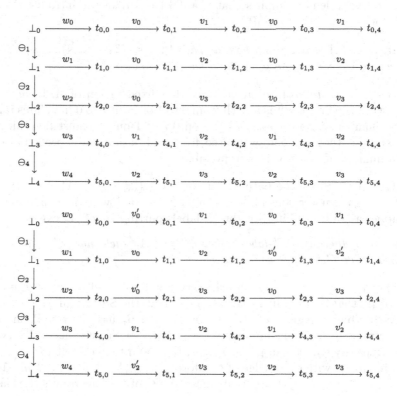

Fig. 4. Top: The TS A_τ as introduced in Sect. 3.1 that originates from Example 3. Bottom: The corresponding Σ'-label-splitting B_τ of A_τ.

3.1 The Proof of Theorem 1.1

In the remainder of this section, unless stated explicitly otherwise, we assume that $\tau = \{\mathsf{nop}, \mathsf{swap}\} \cup \omega$ with is an arbitrary but fixed subset $\omega \subseteq \{\mathsf{inp}, \mathsf{out}, \mathsf{used}, \mathsf{free}\}$.

For a start, we define $\kappa = n + 2m - 1 + \lambda$, where $n + 2m - 1$ is the number of events of the aforementioned TS A_τ. Hence, λ is the maximum number of events of A_τ that could potentially been split in a Σ'-label-splitting B_τ of A_τ.

For every $i \in \{0, \ldots, m-1\}$, the TS A_τ has the following directed path T_i that uses the vertices v_{i_0} and v_{i_1} of the edge $\mathfrak{e}_i$ as events:

$$T_i = t_{i,0} \xrightarrow{v_{i_0}} t_{i,1} \xrightarrow{v_{i_1}} t_{i,2} \xrightarrow{v_{i_0}} t_{i,3} \xrightarrow{v_{i_1}} t_{i,4}$$

Finally, for all $i \in \{0, \ldots, m-1\}$, we apply the edge $\perp_i \xrightarrow{w_i} t_{i,0}$ and, if $i < m-1$, then also the edge $\perp_i \xrightarrow{\ominus_{i+1}} \perp_{i+1}$ to join the paths $T_0, \ldots, T_{m-1}$ into the TS A_τ, cf. Fig. 4. Let $\perp = \{\perp_0, \ldots, \perp_{m-1}\}$ and $W = \{w_0, \ldots, w_{m-1}\}$ and $\ominus = \{\ominus_1, \ldots, \ominus_{m-1}\}$. The TS A_τ has exactly $|V \cup W \cup \ominus| = n + 2m - 1$ events.

The following lemma implies that if a TS has any of the introduced paths, then it does not have the τ-SSP:

Lemma 2. *Let A be a TS that has the path $P_0 = s_0 \xrightarrow{a} s_1 \xrightarrow{b} s_2 \xrightarrow{a} s_3 \xrightarrow{b} s_4$. If $R = (sup, sig)$ is a τ-region of A, then $sup(s_0) = sup(s_4)$.*

Proof. Let $R = (sup, sig)$ be an arbitrary but fixed region of A. If $sup(s_0) \neq sup(s_4)$, then the image P^R is a path from 0 to 1 or from 1 to 0 in τ. This implies that the number of state changes between 0 and 1 on P^R must be odd. Since $sig(a), sig(b) \in \{\mathsf{nop, inp, out, swap, used, free}\}$ and both a and b occur twice, i.e. an even number of times, this is impossible. $\qquad\square$

In fact, if a TS has the path T_i for some $i \in \{0, \ldots, m-1\}$, then the SSP atom $(t_{i,0}, t_{i,4})$ is not τ-solvable by Lemma 2. The following lemma states that this implies a λ-VC of G if a sought Σ'-label-splitting B_τ of A_τ exists:

Lemma 3. *If there is a Σ'-label-splitting B_τ of A_τ such that $|\Sigma'| \leq \kappa$ that has the τ-SSP, then G has a λ-vertex cover.*

Proof. Let $i \in \{0, \ldots, m-1\}$ be arbitrary but fixed, and let $\mathfrak{S}$ be the set of events of A_τ that occur split in B_τ. By Lemma 2, the SSP atom $\alpha_i = (t_{i,0}, t_{i,4})$ is not τ-solvable by regions of A_τ. However, since B_τ has the τ-SSP, the atom α_i is τ-solvable in B_τ. This implies $\{v_{i_0}, v_{i_1}\} \cap \mathfrak{S} \neq \emptyset$. Since i was arbitrary, this is simultaneously true for all paths $T_0, \ldots, T_{m-1}$ and thus the set $M = \{v \in V \mid v \in \mathfrak{S}\}$ intersects with every edge of G. Moreover, by $|\Sigma'| \leq \kappa = n + 2(m-1) + \lambda$, we have $|M| \leq |\mathfrak{S}| \leq \kappa$. Hence, M defines a λ-VC of G. This proves the claim.$\square$

Conversely, let $M = \{v_{j_0}, \ldots, v_{j_{\lambda-1}}\} \subseteq V$ be a λ-VC of G. In the remainder of this section, we argue that there is a sought Σ'-label-splitting B_τ of A_τ. For every $i \in \{0, \ldots, \lambda-1\}$, we split the event v_{j_i} into the two events v_{j_i} and v'_{j_i}. This yields $\Sigma' = (\Sigma \setminus M) \cup \bigcup_{i=0}^{\lambda-1} \{v_{j_i}, v'_{j_i}\}$. To define the aforementioned Σ'-label-splitting $B_\tau = (S, \Sigma', \delta', \perp_0)$ of A_τ, it suffices to define δ' on the states of $T_0, \ldots, T_{m-1}$. In particular, for all $i \in \{0, \ldots, m-1\}$, δ' restricted to $S(T_i)$ and $\Sigma(T_i)$ yields the path T'_i as follows:

- if $v_{i_0} \in M$ and $v_{i_1} \notin M$, then $T'_i = t_{i,0} \xrightarrow{v'_{i_0}} t_{i,1} \xrightarrow{v_{i_1}} t_{i,2}, \xrightarrow{v_{i_0}} t_{i,3} \xrightarrow{v_{i_1}} t_{i,4}$;

- if $v_{i_0}, v_{i_1} \in M$, then $T'_i = t_{i,0} \xrightarrow{v_{i_0}} t_{i,1} \xrightarrow{v_{i_1}} t_{i,2}, \xrightarrow{v'_{i_0}} t_{i,3} \xrightarrow{v'_{i_1}} t_{i,4}$;

- if $v_{i_0} \notin M$ and $v_{i_1} \in M$, then $T'_i = t_{i,0} \xrightarrow{v_{i_0}} t_{i,1} \xrightarrow{v_{i_1}} t_{i,2}, \xrightarrow{v_{i_0}} t_{i,3} \xrightarrow{v'_{i_1}} t_{i,4}$.

The following lemma essentially states that if $sup : S(B_\tau) \to \{0,1\}$ and $sig : \Sigma(B_\tau) \to \tau$ are mappings that define regions when restricted to $T'_0, \ldots, T'_{m-1}$, then they can be extended suitably to a region of B_τ:

Lemma 4. *If* $sup : S(B_\tau) \setminus \bot \to \{0,1\}$ *and* $sig : \Sigma' \setminus (W \cup \ominus) \to \tau$ *are mappings such that* $s \xrightarrow{e} s' \in B_\tau$ *and* $e \notin U$ *implies* $sup(s) \xrightarrow{sig(e)} sup(s') \in \tau$, *then there is a τ-region* $R = (sup', sig')$ *of B_τ that preserves sup and sig as follows:*

1. *For all* $s \in S(B_\tau)$, *if* $s \notin \bot$, *then* $sup'(s) = sup(s)$, *otherwise* $sup'(s) = 0$.
2. *For all* $e \in \Sigma'$, *if* $e \notin W \cup \ominus$, *then* $sig'(e) = sig(e)$; *if* $e \in \ominus$, *then* $sig(e) = \mathsf{nop}$; *if* $i \in \{0, \ldots, m-1\}$ *and* $e = w_i$ *and* $sup(t_{i,0}) = 0$, *then* $sig(e) = \mathsf{nop}$; *otherwise* $sig(e) = \mathsf{swap}$.

Proof. We argue that $s \xrightarrow{e} s' \in B_\tau$ implies $sup'(s) \xrightarrow{sig'(e)} sup'(s') \in \tau$. If $e \in W \cup \ominus$, then this is easy to see, since $sup'(\bot_i) = 0$ for all $i \in \{0, \ldots, m-1\}$. For $e \notin W \cup \ominus$, the claim follows by the assumptions about sup and sig. $\qquad\square$

By the next lemma, a τ-region of T'_i, where T'_i is considered as a TS, whose signature gets along with nop and swap is always extendable to a region of B_τ:

Lemma 5. *Let* $i \in \{0, \ldots, m-1\}$. *Let* $sup : S(T'_i) \to \{0,1\}$ *and* $sig : \Sigma'(T'_i) \to \{\mathsf{nop}, \mathsf{swap}\}$ *be mappings such that* $s \xrightarrow{e} s' \in T'_i$ *implies* $sup(s) \xrightarrow{sig(e)} sup(s') \in \tau$. *There is a τ-region* $R = (sup', sig')$ *of B_τ such that* $sup'(s) = sup(s)$ *and* $sig'(e) = sig(e)$ *for all* $s \in S(T'_i)$ *and* $e \in \Sigma'(T'_i)$.

Proof. By Lemma 4, it suffices to argue that sup and sig are consistently extendable to $T'_0, \ldots T'_{i-1}, T'_{i+1}, \ldots, T'_{m-1}$. Let $j \in \{0, \ldots, m-1\} \setminus \{i\}$, be arbitrary but fixed, and let $T'_j = t_{j,0} \xrightarrow{e_{j,1}} t_{j,1} \xrightarrow{e_{j,2}} t_{j,2} \xrightarrow{e_{j,3}} t_{j,3} \xrightarrow{e_{j,4}} t_{j,4}$, where $e_{j,1}, \ldots, e_{j,4} \in \Sigma'(T'_j)$ in accordance to the definition of B_τ. We obtain $R = (sup', sig')$ as follows. For all $e \in \Sigma'(B_\tau) \setminus (W \cup \ominus)$, if $e \in \Sigma'(T'_i)$, then $sig'(e) = sig(e)$ and otherwise $sig'(e) = \mathsf{nop}$; for all $s \in S(T'_i)$, we define $sup'(s) = sup(s)$; for all $j \in \{0, \ldots, m-1\} \setminus \{i\}$, we define $sup(t_{j,0}) = 0$ and inductively $sup(t_{j,\ell}) = \delta_\tau(sup(t_{j,\ell}), e_{j,\ell})$ for all $\ell \in \{1, \ldots, 4\}$. Since sig maps to $\{\mathsf{nop}, \mathsf{swap}\}$, so does sig'. Thus, if $s \xrightarrow{e} s' \in T'_j$, then $sup'(s) \xrightarrow{sig'(e)} sup'(s') \in \tau$. Since j was arbitrary, this proves the lemma. $\qquad\square$

Lemma 6. *The TS B_τ has the τ-SSP.*

Proof. It is easy to see that $(\bot_i, s)$ is τ-solvable for all $i \in \{0, \ldots, m-1\}$ and all $s \in S(B_\tau) \setminus \{\bot_i\}$. Similarly, the atom (s, s') where $s \in S(T'_i)$ and $s' \in S(B_\tau) \setminus S(T'_i)$ is τ-solvable for all $i \in \{0, \ldots, m-1\}$. Thus, it remains to argue that an atom (s, s') is also solvable if $s \neq s \in S(T'_i)$, for all $i \in \{0, \ldots, m-1\}$. By Lemma 5, it suffices to present corresponding regions for T'_i.

Let $i \in \{0, \ldots, m-1\}$ be arbitrary but fixed, and, for a start, let's consider the case where $v_{i_0} \in M$ and $v_{i_1} \notin M$. That is, $T'_i = t_{i,0} \xrightarrow{v'_{i_0}} t_{i,1} \xrightarrow{v_{i_1}} t_{i,2} \xrightarrow{v_{i_0}} t_{i,2} \xrightarrow{v_{i_1}} t_{i,4}$. For all $\ell \in \{1, 2, 3\}$, let $R_\ell = (sup_\ell, sig_\ell)$ be a pair of mappings $sup_\ell : S(T'_i) \to$

$\{0,1\}$, $sig_\ell : \Sigma'(T_i') \to \{\mathsf{nop}, \mathsf{swap}\}$, (implicitly) defined by $sup_1(t_{i,0}) = 0$, $sig_1(v_{i_0}') = sig_1(v_{i_0}) = \mathsf{nop}$ and $sig_1(v_{i_1}) = \mathsf{swap}$; and $sup_2(t_{i,0}) = 0$, $sig_2(v_{i_0}') = sig_2(v_{i_1}) = \mathsf{nop}$ and $sig_2(v_{i_0}) = \mathsf{swap}$; and $sup_3(t_{i,0}) = 0$, $sig_3(v_{i_0}) = sig_3(v_{i_1}) = \mathsf{nop}$ and $sig_3(v_{i_0}') = \mathsf{swap}$. The images of T_i' under R_1, R_2 and R_3 are as follows:

$$T_i'^{R_1} = 0\xrightarrow{\mathsf{nop}}0\xrightarrow{\mathsf{swap}}1\xrightarrow{\mathsf{nop}}0\xrightarrow{\mathsf{swap}}1 \qquad T_i'^{R_2} = 0\xrightarrow{\mathsf{nop}}0\xrightarrow{\mathsf{nop}}0\xrightarrow{\mathsf{swap}}1\xrightarrow{\mathsf{nop}}1$$

$$T_i'^{R_3} = 0\xrightarrow{\mathsf{swap}}1\xrightarrow{\mathsf{nop}}1\xrightarrow{\mathsf{nop}}1\xrightarrow{\mathsf{nop}}1$$

By Lemma 5, R_ℓ can be extended to a τ-region of B_τ that preserves sup for all $\ell \in \{1,2,3\}$. Moreover, obviously, for every SSP atom (s, s') of T_i', there is an $\ell \in \{1,2,3\}$ such that $sup_\ell(s) \neq sup_\ell(s')$. Thus, (s, s') is τ-solvable in B_τ.

The arguments for the cases $v_{i_0} \notin M$ and $v_{i_1} \in M$ and $v_{i_0} \in M$ and $v_{i_1} \in M$ are similar. By the arbitrariness of i, this proves the lemma. $\square$

3.2   The Proof of Theorem 1.2, Where $\tau \cap \{\mathsf{inp}, \mathsf{out}\} \neq \emptyset$

Let $\tau = \{\mathsf{nop}, \mathsf{swap}\} \cup \omega$ be a type of nets such that $\omega \subseteq \{\mathsf{inp}, \mathsf{out}, \mathsf{used}, \mathsf{free}\}$ and $\omega \cap \{\mathsf{inp}, \mathsf{out}\} \neq \emptyset$, and let A_τ and κ be defined as in Sect. 3.1.

Lemma 7. *If there is a Σ'-label-splitting B_τ of A_τ such that $|\Sigma'| \leq \kappa$ that has the τ-ESSP, then G has a λ-vertex cover.*

Proof. Let $i \in \{0, \ldots, m-1\}$ be arbitrary but fixed and $\mathfrak{S}$ be the set of events of A_τ that occur split in B_τ. If $R = (sup, sig)$ is a τ-region of A_τ, then $sup(t_{i,0}) = sup(t_{i,4})$ by Lemma 2. Thus, $\alpha_i = (v_{i_0}, t_{i,4})$ is not τ-solvable, since $sup(t_{i,0}) \xrightarrow{sig(v_{i_0})}$ implies $sup(t_{i,4}) \xrightarrow{sig(v_{i_0})}$. On the other hand, B_τ has the τ-ESSP, implying the τ-solvability of α_i. This implies $\{v_{i_0}, v_{i_1}\} \cap \mathfrak{S} \neq \emptyset$. Since i was arbitrary, this is true for all $T_0, \ldots, T_{m-1}$. Thus, just like for Lemma 3, we get that $M = \{v \in V \mid v \in \mathfrak{S}\}$ defines a λ-VC of G. This proves the lemma. $\square$

Conversely, let M be a λ-VC of G, and B_τ be the Σ'-label-splitting of A_τ as defined in Sect. 3.1. By the following lemma, B_τ has an exact net realization:

Lemma 8. *The TS B_τ has the τ-SSP and the τ-ESSP.*

Proof. By Lemma 6, the TS B_τ has the τ-SSP. It remains to argue for the τ-ESSP. Without loss of generality, we assume that $\mathsf{inp} \in \tau$ and present τ-regions $R = (sup, sig)$ that get along with $\mathsf{nop}, \mathsf{inp}$ and swap. If $\mathsf{inp} \notin \tau$, then $\mathsf{out} \in \tau$ and one gets corresponding (complement) regions $R' = (sup, sig')$ simply by $sup'(s) = 1 - sup(s)$, and $sig'(e) = sig(e)$ if $sig(e) \in \{\mathsf{nop}, \mathsf{swap}\}$, and $sig'(e) = \mathsf{out}$ if $sig(e) = \mathsf{inp}$ for all $s \in S(B_\tau)$ and all $e \in \Sigma'$.

We proceed as follows. Let $i \in \{0, \ldots, m-1\}$ be arbitrary but fixed. Firstly, we argue that if $e \in \Sigma'(T_i')$ and $s \in S(T_i')$ such that (e, s) is an ESSP atom, then (e, s) is τ-solvable. Secondly, we argue that if $e \in \Sigma'(T_i')$ and $s \in S(B_\tau) \setminus S(T_i')$, then (e, s) is also solvable. Since i was arbitrary, this proves that all ESSP atoms (e, s), where $e \in \Sigma'(B_\tau) \setminus (W \cup \ominus)$ and $s \in S(B_\tau) \setminus \bot$, are τ-solvable.

Finally, we argue that the remaining ESSP atoms are also τ-solvable. For the sake of simplicity, we define all regions $R = (sup, sig)$ implicitly by $sup(\perp_0)$ and sig.

First of all, let's consider the case $v_{i_0} \notin M$ and $v_{i_1} \in M$, which implies

$$T_i' = t_{i,0} \xrightarrow{v_{i_0}} t_{i,1} \xrightarrow{v_{i_1}} t_{i,2}, \xrightarrow{v_{i_0}} t_{i,3} \xrightarrow{v_{i_1}'} t_{i,4}.$$

(v_{i_0}): Let $j \neq \ell \in \{0, \dots, m-1\} \setminus \{i\}$ select the other edges of v_{i_0} in G, i.e., $v_{i_0} \in e_j \cap e_\ell$. Assume that $j_0 < i_0 < \ell_1$, i.e., $v_{j_1} = v_{i_0}$ and $v_{\ell_0} = v_{i_0}$ (the cases $j_0, \ell_0 < i_0$ and $i_0 < j_1, \ell_1$ are similar).

The following τ-region $R_1 = (sup, sig)$ solves (v_{i_0}, s) for all $s \in \{t_{i,1}, t_{i,3}, t_{i,4}\}$: $sup(\iota) = 0$; for all $e \in \Sigma'$, if $e = v_{i_0}$, then $sig(e) = $ inp; if $e \in \{v_{i_1}, v_{j_0}, v_{\ell_1}\} \cup \{w_i, w_j, w_\ell\}$, then $sig(e) = $ swap, otherwise $sig(e) = $ nop. The image $T_i'^{R_1}$ is $1 \xrightarrow{\text{inp}} 0 \xrightarrow{\text{swap}} 1 \xrightarrow{\text{inp}} 0 \xrightarrow{\text{nop}} 0$.

$(v_{i_1}$ and $v_{i_1}')$: Let $j \neq \ell \in \{0, \dots, m-1\} \setminus \{i\}$ such that $v_{i_1} \in e_j \cap e_\ell$. The following τ-region $R_2 = (sup, sig)$ solves (v_{i_1}, s) for all $s \in \{t_{i,0}, t_{i,2}\}$: $sup(\iota) = 0$; for all $e \in \Sigma'$, if $e = v_{i_1}$, then $sig(e) = $ inp; if $e \in \{v_{i_0}\} \cup \{w_j, w_\ell\}$, then $sig(e) = $ swap, otherwise $sig(e) = $ nop. The image $T_i'^{R_2}$ is $0 \xrightarrow{\text{swap}} 1 \xrightarrow{\text{inp}} 0 \xrightarrow{\text{swap}} 1 \xrightarrow{\text{nop}} 1$.

The following τ-region $R_3 = (sup, sig)$ solves (v_{i_1}, s) for all $s \in \{t_{i,3}, t_{i,4}\}$: $sup(\iota) = 0$; for all $e \in \Sigma'$, if $e = v_{i_1}$, then $sig(e) = $ inp; if $e \in \{w_i, w_j, w_\ell\}$, then $sig(e) = $ swap, otherwise $sig(e) = $ nop. It is $T_i'^{R_3} = 1 \xrightarrow{\text{nop}} 1 \xrightarrow{\text{inp}} 0 \xrightarrow{\text{nop}} 0 \xrightarrow{\text{nop}} 0$.

The following τ-region $R_4 = (sup, sig)$ solves (v_{i_1}', s) for all $s \in \{t_{i,1}, t_{i,2}, t_{i,4}\}$: $sup(\iota) = 0$; for all $e \in \Sigma'$, if $e = v_{i_1}'$, then $sig(e) = $ inp; if $e \in \{v_{i_0}\} \cup \{w_i, w_j, w_\ell\}$, then $sig(e) = $ swap, otherwise $sig(e) = $ nop; we have $T_i'^{R_4} = 1 \xrightarrow{\text{swap}} 0 \xrightarrow{\text{nop}} 0 \xrightarrow{\text{swap}} 1 \xrightarrow{\text{inp}} 0$.

The following τ-region $R_5 = (sup, sig)$ solves $(v_{i_1}', t_{i,0})$: $sup(\iota) = 0$; for all $e \in \Sigma'$, if $e = v_{i_1}'$, then $sig(e) = $ inp; if $e = v_{i_1}$, then $sig(e) = $ swap; if $e = w_j$ and $v_{j_0} = v_{i_1}$, implying $v_{i_1}' v_{j_1} v_{i_1} v_{j_1}$ in T_j, then $sig(e) = $ swap; if $e = w_\ell$ and $v_{\ell_0} = v_{i_1}$, implying $v_{i_1}' v_{\ell_1} v_{i_1} v_{\ell_0}$ in T_ℓ, then $sig(e) = $ swap; otherwise $sig(e) = $ nop; The image $T_i'^{R_5}$ is $0 \xrightarrow{\text{nop}} 0 \xrightarrow{\text{swap}} 1 \xrightarrow{\text{nop}} 1 \xrightarrow{\text{inp}} 0$.

One finds out that the case $v_{i_0} \in M$ and $v_{i_1} \notin M$ is similar to the just presented one, and the case $v_{i_0}, v_{i_1} \in M$ is even simpler. The latter is due to the fact that if $e \in \{v_{i_0}, v_{i_1}, v_{i_0}', v_{i_1}'\}$ and $j \in \{0, \dots, m-1\}$ such that $e \in \Sigma'(T_j')$, then e occurs exactly once in T_j'. For the sake of simplicity, we refrain from the exhaustive case analyses. Notice that, by the arbitrariness of i, the presented regions also solve (e, s) for all $e \in V \cup M$ and $s \in \perp$, since they map e to inp and s to 0.

We argue that, for all $i \in \{0, \dots, n-1\}$ and $j \in \{0, \dots, m-1\}$, if $e \in \{v_i, v_i'\}$, $s \in T_j'$ and $e \notin \Sigma'(T_j')$, then (e, s) is τ-solvable: If $e \in \{v_i, v_i'\}$, then there is a τ-region $R \in \{R_1, \dots, R_5\}$ such that $sup(\perp_0) = 0$, $sig(e) = $ inp and $sig(e') \in \{$nop, swap$\}$ for all $e' \Sigma'(B_\tau) \setminus \{e\}$. Let $j \in \{0, \dots, m-1\}$ such that $e \notin \Sigma(T_j')$, which implies $sig(w_j) = $ nop. Let $s \in S(T_j')$. If $sup(s) = 0$, then (e, s) is solved. Otherwise, if $sup(s) = 1$, then s is preceded by an event whose signature is swap. Thus, we simply modify R to a corresponding region R' by mapping w_j to swap instead of nop. R' is well defined, since all events of T_j' are mapped to nop or swap, and solves (e, s). Finally, it is easy to see that, for all $e \in W \cup \ominus$ and $s \in S(B_\tau)$ such that $\neg s \xrightarrow{e}$, the atom (e, s) is τ-solvable. This proves the lemma. □

3.3 The Proof of Theorem 1.2, Where $\tau \cap \{\mathsf{inp}, \mathsf{out}\} = \emptyset$

In the remainder of this section, unless stated explicitly otherwise, let $\tau = \{\mathsf{nop}, \mathsf{swap}\} \cup \omega$ be a type of net such that $\emptyset \neq \omega \subseteq \{\mathsf{used}, \mathsf{free}\}$.

Notice that, for any TS A, if $s \xrightarrow{e} s' \in A$ and $\neg s' \xrightarrow{e}$, then (e, s') is not τ-solvable. Thus, there is no Σ'-label-splitting of the TS A_τ of Sect. 3.1, that has the τ-ESSP. To overcome this obstacle, with as little effort as possible, the current TS A_τ extends the one of Sect. 3.1 by backward-edges: For all $i \in \{0, \ldots, m-1\}$, the current TS A_τ has the following bi-directed path T_i:

$$T_i = t_{i,0} \overset{v_{i_0}}{\longleftrightarrow} t_{i,1} \overset{v_{i_1}}{\longleftrightarrow} t_{i,2} \overset{v_{i_0}}{\longleftrightarrow} t_{i,3} \overset{v_{i_1}}{\longleftrightarrow} t_{i,4}$$

Moreover, the TS A_τ has, for all $i \in \{0, \ldots, m-1\}$, the bi-directed edge $\perp_i \overset{w_i}{\longleftrightarrow} t_{i,0}$; and, for all $i \in \{0, \ldots, m-2\}$, the bi-directed edge $\perp_i \overset{\ominus_{i+1}}{\longleftrightarrow} \perp_{i+1}$ to join the paths $T_0, \ldots, T_{m-1}$. Again, A_τ has $n + 2m - 1$ events, and we define $\kappa = n + 2m - 1 + \lambda$.

Lemma 9. *If there is a Σ'-label-splitting C_τ of A_τ such that $|\Sigma'| \leq \kappa$ that has the τ-ESSP, then G has a λ-vertex cover.*

Proof. The proof is similar to the one of Lemma 7. □

Conversely, let M be a λ-VC of G, and let $C_\tau = (S(B_\tau), \Sigma', \delta'', \perp_0)$ be the bi-directed extension of the TS $B_\tau = (S(B_\tau), \Sigma', \delta', \perp_0)$, which has been defined in Sect. 3.1. That is, for all $s, s' \in S(B_\tau)$ and all $e \in \Sigma'$ if $\delta'(s, e) = s'$, then $\delta''(s, e) = s'$ and $\delta''(s', e) = s$. To complete the proof of Theorem 1.2 it remains to argue that C_τ has the τ-ESSP and the τ-SSP. Recall that the signatures of the regions that have been presented for the proof of Lemma 6 get along with nop and swap. Thus, they can be directly applied to C_τ, which proves C_τ's τ-SSP. Hence, it remains to argue for C_τ's ESSP. The following lemma confirms both properties for C_τ and thus completes the proof of Theorem 1.

Lemma 10. *The TS C_τ has the τ-SSP and the τ-ESSP.*

Proof. We argue that C_τ has the τ-ESSP. Without loss of generality, we assume used $\in \tau$. (Otherwise free $\in \tau$ and this case is similar.) Recall that G is 3-bounded, that is, every node-event or its splitting occur at most three times in C_τ. For a start, it is easy to see that if $e \in \perp \cup W$, then (e, s) is τ-solvable for all relevant $s \in S(C_\tau)$.

Let $i \in \{0, \ldots, m-1\}$ and $e \in \Sigma'(T_i)$ be arbitrary but fixed, and let $j \neq \ell \in \{0, \ldots, m-1\} \setminus \{i\}$ such that $e \in \Sigma'(T_j') \cap \Sigma'(T_\ell')$ (if they exist). The following region $R = (sup, sig)$ solves (e, s) for all $s \in S(C_\tau) \setminus (S(T_i') \cup S(T_j') \cup S(T_\ell'))$:

$sup(\bot_0) = 0$; for all $e' \in \Sigma'$, if $e' = e$, then $sig(e) = $ used; if $e' \in \{w_i, w_j, w_\ell\}$, then $sig(e') = $ swap, otherwise, $sig(e') = $ nop.

Thus, it remains to argue that (e, s) is also solvable is e and s occur in the same T_i'. To do so, we let $i \in \{0, \ldots, m-1\}$ and start with he case where $v_{i_0} \in M$ and $v_{i_1} \notin M$, that is, $T_i' = t_{i,0} \xleftarrow{v_{i_0}'} t_{i,1} \xrightarrow{v_{i_1}} t_{i,2} \xleftarrow{v_{i_0}} t_{i,3} \xrightarrow{v_{i_1}} t_{i,4}$, and show (e, s) is τ-solvable for all $e \in \{v_{i_0}, v_{i_0}', v_{i_1}\}$ and all $s \in S(T_i')$. Recall that v_{i_0} and v_{i_1} unambiguously define an edge of the graph G, that is, they never occur bot in two distinct $T_i' \neq T_j'$. Moreover, by definition of C_τ, a prime-event v' does never occur between an un-split event $u \in V$, that is, $uv'u$ is not possible.

(v_{i_0}'): Let $j \neq \ell \in \{0, \ldots, m-1\} \setminus \{i\}$, such that $v_{i_0}' \in \Sigma'(T_j') \cap \Sigma'(T_\ell')$. Notice that this implies that $v_{i_1} \notin \Sigma'(T_j') \cup \Sigma'(T_\ell')$.

First of all, we want to define a τ-region $R = (sup, sig)$ that solves (v_{i_0}', s) for all $s \in \{t_{i,2}, t_{i,3}\}$ such that $T_i'^R = 1 \xrightarrow{\text{used}} 1 \xrightarrow{\text{swap}} 0 \xrightarrow{\text{nop}} 0 \xrightarrow{\text{swap}} 1$. The following definition is sound and satisfies this requirement: $sup(\bot_0) = 0$; for all $e \in \Sigma'$, if $e = v_{i_0}$, then $sig(e) = $ used; if $e \in \{w_i, w_j, w_\ell, v_{i_1}\}$, then $sig(e) = $ swap; otherwise, $sig(e) = $ nop.

To solve the remaining atom $(v_{i_0}', t_{i,4})$, we want to define a τ-region $R = (sup, sig)$ such that $T_i'^R = 1 \xrightarrow{\text{used}} 1 \xrightarrow{\text{nop}} 1 \xrightarrow{\text{swap}} 0 \xrightarrow{\text{nop}} 0$. Up to isomorphism or renaming, two situations can occur for the shape of T_j' and T_ℓ':

$$T_j' = t_{j,0} \xleftarrow{v_{i_0}'} t_{j,1} \xrightarrow{v_{j_1}} t_{j,2} \xleftarrow{v_{i_0}} t_{j,3} \xrightarrow{v_{j_1}} t_{j,4},$$

$$T_\ell' = t_{\ell,0} \xleftarrow{v_{\ell_0}} t_{\ell,1} \xrightarrow{v_{i_0}} t_{\ell,2} \xleftarrow{v_{\ell_0}} t_{\ell,3} \xrightarrow{v_{i_0}'} t_{\ell,4}.$$

To define R appropriately, we want to ensure $T_j'^R = 1 \xrightarrow{\text{used}} 1 \xrightarrow{\text{nop}} 1 \xrightarrow{\text{swap}} 0 \xrightarrow{\text{nop}} 0$ and $T_\ell'^R = 0 \xrightarrow{\text{nop}} 0 \xrightarrow{\text{swap}} 1 \xrightarrow{\text{nop}} 1 \xrightarrow{\text{used}} 1$. The following definition is sound and satisfies this requirement: $sup(\bot_0) = 0$; for all $e \in \Sigma'$, if $e = v_{i_0}$, then $sig(e) = $ used; if $e \in \{w_i, w_j, v_{i_0}\}$, then $sig(e) = $ swap; otherwise, $sig(e) = $ nop.

(v_{i_0}): The next region $R = (sup, sig)$ yields $T_i'^R = 0 \xrightarrow{\text{nop}} 0 \xrightarrow{\text{swap}} 1 \xrightarrow{\text{used}} 1 \xrightarrow{\text{swap}} 0$ and thus solves (v_{i_0}, s) for all $s \in \{t_{i,0}, t_{i,1}, t_{i,4}\}$: $sup(\bot_0) = 0$; for all $e \in \Sigma'$, if $e = v_{i_0}$, then $sig(e) = $ used; if $e \in \{w_j, w_\ell, v_{i_1}\}$, then $sig(e) = $ swap; otherwise $sig(e) = $ nop.

(v_{i_1}): The next region $R = (sup, sig)$ yields $T_i'^R = 0 \xrightarrow{\text{swap}} 1 \xrightarrow{\text{used}} 1 \xrightarrow{\text{nop}} 1 \xrightarrow{\text{used}} 1$ and thus solves $(v_{i_1}, t_{i,0})$: $sup(\bot_0) = 1$; for all $e \in \Sigma'$, if $e = v_{i_1}$, then $sig(e) = $ used; if $e \in \{w_i, v_{i_1}\}$, then $sig(e) = $ swap; otherwise $sig(e) = $ nop.

Altogether, by the arbitrariness if i, we have shown that if $v_{i_0} \in M$ and $v_{i_1} \notin M$, then all well-defined ESSP atoms (e, s) such that $e \in \Sigma'(T_i')$ and $s \in S(T_i')$ are τ-solvable for all $i \in \{0, \ldots, m-1\}$. One finds out that the cases where $v_{i_0} \notin M$ and $v_{i_1} \in M$ as well as $v_{i_0}, v_{i_1} \in M$ can be solved similar. This finally proves the τ-ESSP for C_τ and thus completes the proof of Lemma 10. $\square$

4 Concluding Remarks

In this paper, we characterize the computational complexity of finding a label-splitting of a TS A that allows an implementing τ-net for all types $\tau = \{\mathsf{nop}, \mathsf{swap}\} \cup \omega$ with $\omega \subseteq \{\mathsf{inp}, \mathsf{out}, \mathsf{used}, \mathsf{free}\}$ and all implementations previously considered in the literature. By the results of [19,22,23], the synthesis problem aiming at language-simulation and realization is NP-complete for all types $\tau = \{\mathsf{nop}, \mathsf{swap}\} \cup$ with $\omega \subseteq \{\mathsf{inp}, \mathsf{out}, \mathsf{res}, \mathsf{set}, \mathsf{used}, \mathsf{free}\}$ and $\omega \cap \{\mathsf{inp}, \mathsf{out}, \mathsf{used}, \mathsf{free}\} \neq \emptyset$ and $\omega \cap \{\mathsf{set}, \mathsf{res}\} \neq \emptyset$. Hence, their corresponding label splitting problem is NP-complete. Moreover, similar to the proof of Theorem 1, one argues that label splitting aiming at language simulation or realization is trivial for $\tau = \{\mathsf{nop}, \mathsf{swap}\} \cup \omega$ for all $\omega \subseteq \{\mathsf{res}, \mathsf{set}\}$. Moreover, we already know that synthesis aiming at embedding is NP-complete for all Boolean types τ with $\{\mathsf{nop}, \mathsf{swap}\} \subseteq \tau$ and $\tau \cap \{\mathsf{res}, \mathsf{set}\} = \emptyset$ [21]. Hence, again their corresponding label splitting problem is also NP-complete. Altogether, with the present work, the complexity of τ-label-splitting is characterized for all 64 Boolean types with $\{\mathsf{nop}, \mathsf{swap}\} \subseteq \tau$ and all implementations.

It remains future work to determine the complexity of τ-label-splitting for the swap-free and nop-equipped Boolean types whose underlying synthesis problem is in P. Moreover, one might consider κ as a parameter and investigate the label-splitting problem from the point of view of parameterized complexity. Certainly, the first question to solve in this context is whether the label splitting problem is even in the complexity class XP.

Acknowledgements. I would like to thank the unknown reviewers for their valuable comments.

References

1. van der Aalst, W.M.P.: Process Mining - Discovery, Conformance and Enhancement of Business Processes. Springer, Heidelberg (2011). https://doi.org/10.1007/978-3-642-19345-3
2. Badouel, E., Bernardinello, L., Darondeau, P.: Polynomial algorithms for the synthesis of bounded nets. In: Mosses, P.D., Nielsen, M., Schwartzbach, M.I. (eds.) CAAP 1995. LNCS, vol. 915, pp. 364–378. Springer, Heidelberg (1995). https://doi.org/10.1007/3-540-59293-8_207
3. Badouel, E., Bernardinello, L., Darondeau, P.: The synthesis problem for elementary net systems is NP-complete. Theoret. Comput. Sci. **186**(1–2), 107–134 (1997). https://doi.org/10.1016/S0304-3975(96)00219-8
4. Badouel, E., Bernardinello, L., Darondeau, P.: Petri Net Synthesis. TTCSAES. Springer, Heidelberg (2015). https://doi.org/10.1007/978-3-662-47967-4
5. Badouel, E., Caillaud, B., Darondeau, P.: Distributing finite automata through Petri net synthesis. Formal Asp. Comput. **13**(6), 447–470 (2002). https://doi.org/10.1007/s001650200022
6. Badouel, E., Darondeau, P.: Trace nets and process automata. Acta Informatica **32**(7), 647–679 (1995). https://doi.org/10.1007/BF01186645

7. Carmona, J.: The label splitting problem. In: Jensen, K., van der Aalst, W.M., Ajmone Marsan, M., Franceschinis, G., Kleijn, J., Kristensen, L.M. (eds.) Transactions on Petri Nets and Other Models of Concurrency VI. LNCS, vol. 7400, pp. 1–23. Springer, Heidelberg (2012). https://doi.org/10.1007/978-3-642-35179-2_1

8. Cortadella, J., Kishinevsky, M., Lavagno, L., Yakovlev, A.: Deriving petri nets from finite transition systems. IEEE Trans. Comput. **47**(8), 859–882 (1998)

9. Cortadella, J., Kishinevsky, M., Kondratyev, A., Lavagno, L., Yakovlev, A.: A region-based theory for state assignment in speed-independent circuits. IEEE Trans. CAD Integr. Circuits Syst. **16**(8), 793–812 (1997). https://doi.org/10.1109/43.644602

10. Garey, M.R., Johnson, D.S.: Computers and Intractability: A Guide to the Theory of NP-Completeness. W. H. Freeman, New York (1979)

11. Holloway, L.E., Krogh, B.H., Giua, A.: A survey of Petri net methods for controlled discrete event systems. Discrete Event Dyn. Syst. **7**(2), 151–190 (1997). https://doi.org/10.1023/A:1008271916548

12. Kleijn, J., Koutny, M., Pietkiewicz-Koutny, M., Rozenberg, G.: Step semantics of Boolean nets. Acta Informatica **50**(1), 15–39 (2013). https://doi.org/10.1007/s00236-012-0170-2

13. Montanari, U., Rossi, F.: Contextual nets. Acta Informatica **32**(6), 545–596 (1995). https://doi.org/10.1007/BF01178907

14. Pietkiewicz-Koutny, M.: Transition systems of Elementary Net Systems with inhibitor arcs. In: Azéma, P., Balbo, G. (eds.) ICATPN 1997. LNCS, vol. 1248, pp. 310–327. Springer, Heidelberg (1997). https://doi.org/10.1007/3-540-63139-9_43

15. Rozenberg, G., Engelfriet, J.: Elementary net systems. In: Reisig, W., Rozenberg, G. (eds.) ACPN 1996. LNCS, vol. 1491, pp. 12–121. Springer, Heidelberg (1998). https://doi.org/10.1007/3-540-65306-6_14

16. Schlachter, U., Wimmel, H.: Relabelling LTS for Petri net synthesis via solving separation problems. In: Koutny, M., Pomello, L., Kristensen, L.M. (eds.) Transactions on Petri Nets and Other Models of Concurrency XIV. LNCS, vol. 11790, pp. 222–254. Springer, Heidelberg (2019). https://doi.org/10.1007/978-3-662-60651-3_9

17. Schlachter, U., Wimmel, H.: Optimal label splitting for embedding an LTS into an arbitrary Petri net reachability graph is NP-complete. CoRR abs/2002.04841 (2020). https://arxiv.org/abs/2002.04841

18. Schmitt, V.: Flip-flop nets. In: Puech, C., Reischuk, R. (eds.) STACS 1996. LNCS, vol. 1046, pp. 515–528. Springer, Heidelberg (1996). https://doi.org/10.1007/3-540-60922-9_42

19. Tredup, R.: The complexity of synthesizing nop-equipped Boolean nets from g-bounded inputs. Technical report (2019)

20. Tredup, R.: Finding an optimal label-splitting to make a transition system petri net implementable: a complete complexity characterization. In: Italian Conference on Theoretical Computer Science - 21st Annual Conference, ICTCS 2020 (2020, to appear)

21. Tredup, R., Erofeev, E.: The complexity of Boolean state separation (2020). Submitted to ICTAC 2020 (2020)

22. Gopal, T.V., Watada, J. (eds.): TAMC 2019. LNCS, vol. 11436. Springer, Cham (2019). https://doi.org/10.1007/978-3-030-14812-6

23. Tredup, R., Rosenke, C.: On the hardness of synthesizing Boolean nets. In: ATAED@Petri Nets/ACSD. CEUR Workshop Proceedings, vol. 2371, pp. 71–86 (2019). CEUR-WS.org

Author Index

Printed in the United States
By Bookmasters